EARTH'S HIDDEN REALITY

Discover It, Explore It, Embrace It

MARK HUNTER BROOKS

I0541916

SPARK Publications
Charlotte, North Carolina

Earth's Hidden Reality
Discover It, Explore It, Embrace It
By Mark Hunter Brooks

Published by SPARK Publications,
SPARKpublications.com
Charlotte, North Carolina

Cover Design: MaryDes Designs (www.marydes.eu)
Stock Illustration: Glebstock / Adobestock.com, sakkmesterke / Shutterstock.com
Content Editing: Brian Faulkner (www.faulknerwriter.com)
Photo Credits: Unless otherwise noted, all photos were taken by the author
Illustration Credits: Unless otherwise noted, all illustrations were drawn by James Denk, SPARK Publications

Videos highlighting selected topics can be found on the Earth's Hidden Reality YouTube Channel at https://www.youtube.com/channel/UCz52KrH5ytQolGJQb3lfTZA

Paperback, June 2022, ISBN: 978-1-953555-25-0
Library of Congress Control Number: 2022905798

Ebook, August 2022, ISBN: 978-1-953555-26-7

RELIGION / Spirituality
SCIENCE / Waves & Wave Mechanics

The concepts expressed in this book are presented to stimulate thinking and generate conversation about Earth's Hidden Reality. The author's views are not necessarily those of the sources cited. Digital links were current as of the date first published. For specific assistance or information—such as (but not limited to) counseling, advice, diagnoses, options, or treatment—see your physician, lawyer or other professional. Attempting to perform anything described in this book is done at the reader's own discretion. If any adverse effects are encountered, stop immediately and seek professional advice.

Dedication

To Jibril Caliph Al-Sadat,
a great man with a kind heart
I am proud to call a friend

"The decisive question for man is:
Is he related to something
infinite or not?
That is the telling
question of his life."

– C.G. Jung

About the Cover Illustration

The cover drawing is of a spirit I've seen hovering over my bed at night. On one occasion, I put my hand up into its body, and my fingers tingled when they touched it. This spirit is composed of various-sized small balls circulating around a larger central sphere. Actually, there were at least twice as many smaller balls moving around the large sphere than depicted. (I figured having more smaller balls would unnecessarily complicate the cover design, which is why you see what's presented—and the disclaimer!)

If this short description grabs your attention, you've got hold of the right book. There's a lot more about the spiritual world to discover and appreciate as you read—plus the science that supports my belief that the world we see, hear and experience around us every day is wave-based rather than particle-based, as described by classical physics.

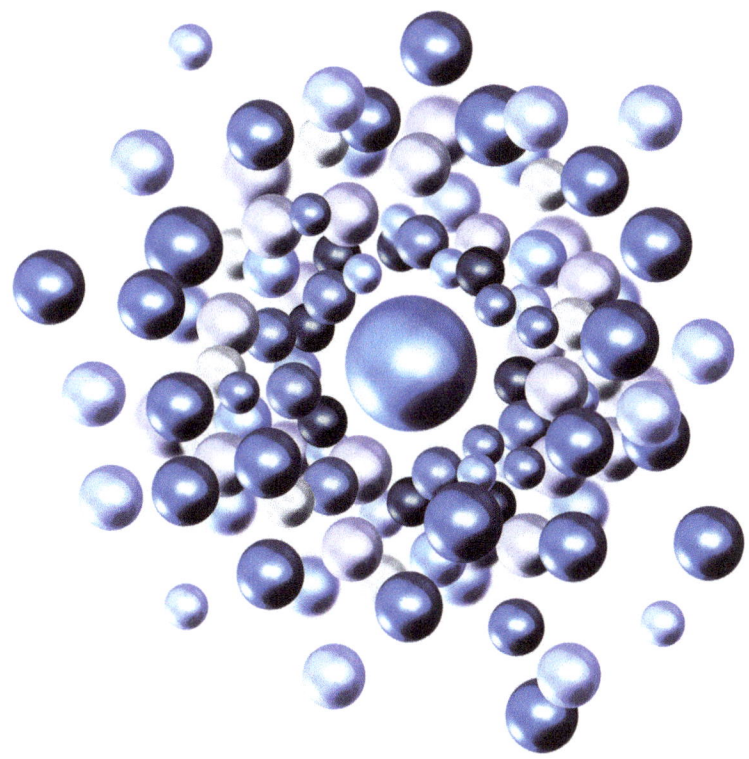

Acknowledgments

My thanks go first to the Bigelow Institute for Consciousness Studies (BICS) for creating their consciousness essay competition. This book began as an essay for that competition, but I withdrew from consideration after realizing that my manuscript was moving beyond its scope. Even so, the BICS competition gave me the impetus to write this book, and for that I am grateful. My explanation of death in chapter seven was part of the essay.

Thanks also go to Brian E. Faulkner, the book's content editor, for the many hours he invested in making my writing easier to understand. Brian is an Emmy Award-winning writer, and I couldn't be in more capable hands! He is very detail-oriented and his determination to complete work on time made me very comfortable working with him.

Special thanks go to Maryna at MaryDes Designs LTD for her cover design and custom illustration work. Maryna designed an award-winning cover for an author friend of mine, and I wanted to "ride the coattails" of that success! I am very pleased with her work.

Lastly, I want to thank the SPARKlers at SPARK Publications who helped put this book together and for creating a beautiful finished product.

- To James Denk, who has illustrated many of the interior page graphics in every book since the first one was published in 2016, for simplifying my ideas and ensuring that they had a consistent design, and
- To founder and Chief SPARKler Fabi Preslar, whose expertise and insight helped guide its development.

TABLE OF CONTENTS

INTRODUCTION **13**

ESSENTIAL CONCEPTS..**15**
THREE STRUGGLES...**16**
STRUGGLE I — DISCOVER IT: BELIEVING THE NON-PHYSICAL WORLD EXISTS **16**
    SCIENCE PARADOXES **17**
    MIRACLES IN THE BIBLE **18**
    PARANORMAL ABILITIES **18**
STRUGGLE II — EXPLORE IT: UNDERSTANDING WHY THE NON-PHYSICAL WORLD EXISTS **18**
STRUGGLE III — EMBRACE IT: MAKING THIS NEW REALITY YOUR OWN **19**
APPENDICES **19**
WHAT THIS BOOK IS NOT ABOUT **20**
CONSIDER THE CORPUS...**20**

ESSENTIAL CONCEPTS **21**

THE THEOSOPHICAL SOCIETY...**21**
ANU AND OCCULT CHEMISTRY...**22**
TORUS...**22**
CHAKRAS..**23**
QUANTUM ENTANGLEMENT..**24**

STRUGGLE I — DISCOVER IT **25**

CHAPTER ONE HOW MY SPIRITUAL EXPERIENCES STARTED **26**
    FURTHER READING – WAVE-BASED MATTER IN HISTORY...............................**28**

CHAPTER TWO THE DOOR TO THE SPIRIT WORLD OPENS.................................**32**
    THE EARLY YEARS (2003-2007)..**32**
    QUESTIONING MY FAITH (2008-2009) **34**
    BUILDING A UNIQUE SPIRITUALITY (2010-2013)...**34**
    BROADENING MY UNDERSTANDING OF OTHER FAITHS (2014-2015)..................**35**
    PUBLISHED WORKS (2016-2019) **36**
    FURTHER READING – DOES ANGULAR MOMENTUM CREATE MASS? **38**

CHAPTER THREE AN INTRODUCTION TO WAVE-BASED DIMENSIONS....................**40**
    HOW WAVE-BASED REALITIES CO-EXIST IN THE SAME SPACE **42**
    HOW OUR CONSCIOUSNESS USES DIFFERENT BODIES TO MOVE BETWEEN DIMENSIONS **42**
    FURTHER READING – QUANTUM DISENTANGLEMENT ON A COSMIC SCALE...........**44**
        COULD ANU BE THE SOURCE OF COSMIC MOTION...................................**46**
        …AND THE COLORS IN THE HUMAN AURA? **47**

CHAPTER FOUR EXPLAINING SCIENCE ANOMALIES AND
    PARADOXES WITH WAVE-BASED PHYSICS **49**
    A THEORY OF EVERYTHING (TOE) **49**
    STRONG AND WEAK ATOMIC FORCES **49**
    MAGNETISM **50**
    DARK MATTER AND DARK ENERGY **51**
    FURTHER READING – HOW DIFFERENT DIMENSIONS
        CAN HAVE THE SAME VIBRATIONAL FREQUENCIES WITHIN THEM................**51**

CHAPTER FIVE EXPLAINING MIRACLES IN THE BIBLE WITH WAVE-BASED PHYSICS...................**54**
    FURTHER READING – A WAVE-BASED EXPLANATION OF THE DOUBLE-SLIT EXPERIMENT **57**

CHAPTER SIX EXPLAINING PARANORMAL ABILITIES WITH WAVE-BASED PHYSICS **58**
    UFO APPEARANCE, SPEED AND MANEUVERABILITY **58**
    UNSEEN ASSISTANCE—INVISIBLE HELPERS **59**
    MEDIUMSHIP AND CHANNELING **60**
    FURTHER READING – PROPHETS, MEDIUMS AND THE ETHICAL USE OF SPIRITUAL ABILITIES ..**60**

CHAPTER SEVEN EXPLAINING DEATH WITH WAVE-BASED PHYSICS **63**

CHAPTER EIGHT DOES INFINITY EXIST IN NATURE? **65**
    INFINITY'S EXISTENCE IN FOUR DIMENSIONS...**65**
    SCALE AS AN EXAMPLE OF INFINITY IN NATURE **66**
    FURTHER READING – DEATH AS A FOURTH-DIMENSIONAL PROCESS **69**

STRUGGLE II — EXPLORE IT **71**

CHAPTER NINE PREPARING FOR SPIRITUAL EXPERIENCES...**72**
    SEEK DIRECTION...**74**
    SPIRITUAL DEVELOPMENT FRAMEWORKS **74**
    TWO PRINCIPLES TO REMEMBER..**76**
    FURTHER READING – MY SPIRITUAL DEVELOPMENT FRAMEWORK—AN OVERVIEW............**76**

CHAPTER TEN FOOLS GO WHERE ANGELS FEAR TO TREAD **79**
    REMOVING THE UNKNOWN...**79**
    DEALING WITH SPIRITUAL ENTITIES **80**
    EMBRACING ENERGY HEALING..**81**
    DEVELOPING ENERGY SENSITIVITY...**81**
    FORCING SPIRITUAL ABILITIES **82**
    FURTHER READING – SPIRITUAL EMERGENCIES—WHERE TO GET ASSISTANCE **82**

CHAPTER ELEVEN SPIRITUAL EVOLUTION **83**
    FURTHER READING – THE POWER OF YOUR THOUGHTS **84**

CHAPTER TWELVE THOUGHTS REGARDING SPIRITUAL ABILITIES.......................................**88**
    FURTHER READING – HEALING OVERVIEW **90**
        HEALING CATEGORIES **91**
        WHAT GETS HEALED...**92**
        WAVE-BASED HEALING DIAGNOSTICS..**93**
        HEALING EXPERIMENTS–VISUAL AND HANDS ON **94**
        BOOKS TO READ...**94**

STRUGGLE III — EMBRACE IT 96

CHAPTER THIRTEEN DREAM BIG! **97**
  FURTHER READING – DEVELOPING A SINGLE GLOBAL WORLDVIEW **97**

CHAPTER FOURTEEN A WAVE-BASED PHYSICS MODEL OF REALITY **99**
  SECTION 1: THE FORMATION OF EARTH'S DIMENSIONS **99**
  SECTION 2: BALANCING OF ENERGY BETWEEN DIMENSIONS **102**
  SECTION 3: THE PROCESS OF BIRTH, DEATH AND LIFE
    BETWEEN LIVES FOR CONSCIOUS ENTITIES **104**
    THE PROCESS OF BIRTH AND DEATH **105**
  SECTION 4: SPIRITUAL ABILITIES CONSCIOUS ENTITIES ACQUIRE IN
    DIFFERENT DIMENSIONS **108**
  FURTHER READING – BOOKS ON THE SPIRITUAL DEVELOPMENT PROCESS **108**

CHAPTER FIFTEEN RECAPPING CLAIMS, CONCEPTS AND IDEAS **109**
  FURTHER READING – YOUR CONSCIOUS, SUBCONSCIOUS AND HIGHER-LEVEL SELF **112**
    WHY DO I BELIEVE THESE THINGS? **113**
  CONCLUSION – THE FUTURE WILL BE DETERMINED BY OUR LOVE FOR ONE ANOTHER........**114**
  WHERE TO FIND ADDITIONAL INFORMATION **116**

APPENDICES 118

APPENDIX A OBSERVATIONS AND EXPERIMENTS TO FALSIFY CLAIMS MADE IN THIS BOOK......**119**
  OBSERVATIONS SUPPORTING THE CLAIM THAT PHYSICS IS WAVE-BASED **119**
  HOW TO FALSIFY THE CLAIM THAT PHYSICS IS WAVE-BASED—
    MEASURE NEUTRON DECAY TIME IN SPACE..**121**
  OBSERVATIONS SUPPORTING THE CLAIM OF A MULTI-DIMENSIONAL REALITY **122**
  HOW TO FALSIFY THE CLAIM THAT MULTIPLE DIMENSIONS EXIST #1—DISCOVER ANU **125**
  HOW TO FALSIFY THE CLAIM THAT MULTIPLE DIMENSIONS EXIST #2—
    SUBATOMIC PARTICLE DESK STUDY **126**
  OBSERVATIONS SUPPORTING THE EXISTENCE OF BEINGS IN WAVE-BASED DIMENSIONS **126**
  HOW TO FALSIFY THE CLAIM THAT BEINGS EXIST IN MULTIPLE
    DIMENSIONS—PHYSICAL MEDIUMSHIP **128**

APPENDIX B WAVE-BASED ATOMIC FORCES...**130**
  MORE JUSTIFICATION FOR USING QUARKS TO MODEL ATOMIC FORCES **134**

APPENDIX C A HIERARCHICAL TORUS STRUCTURE OF THE UNIVERSE **136**

APPENDIX D FRAMEWORK FOR SPIRITUAL DEVELOPMENT **141**
  STEP 1 – ENTERING THE SPIRITUAL PATH **143**
  STEP 2 – DEVELOPING AN INTELLECTUAL UNDERSTANDING OF FAITH **144**
  STEP 3 – RECONCILING YOUR INTELLECTUAL UNDERSTANDING OF
    FAITH WITH YOUR LIFE EXPERIENCES **144**
  STEP 4 – MATURING A UNIQUE SPIRITUALITY **145**
  SPIRITUAL DEVELOPMENT—AN INDIVIDUAL JOURNEY **146**
  PUTTING IT ALL TOGETHER—BUILDING MORAL COURAGE **146**
    WHY DEVELOP MORAL COURAGE? **148**
    CULMINATION OF THE PROCESS—BECOMING SPIRITUALLY SELF-AWARE................**148**

APPENDIX E WAVE-BASED EXPLANATIONS FOR MIRACLES IN BIBLE SCRIPTURE **150**
  TELEPATHY: MENTALLY COMMUNICATING IMAGES OR THOUGHTS
    BETWEEN TWO OR MORE BEINGS...**150**
     UNIVERSAL HEARING – WHERE A GROUP OF PEOPLE HEAR THE SAME
      MESSAGE BUT IN THEIR OWN LANGUAGE..**150**
     SPIRIT COMMUNICATIONS – WHERE A NON-PHYSICAL BEING COMMUNICATES WITH A HUMAN **150**
     SPIRIT COMMUNICATION VIA TRANCE – HUMANS RECEIVING A
      COMMUNICATION THROUGH A TRANCE OR VISION **151**
     SPIRIT COMMUNICATION VIA DREAMS – HUMANS RECEIVING A COMMUNICATION VIA A DREAM...**151**
  REMOTE PERCEPTION: PROJECTING ONE'S CONSCIOUS
    ENERGY TO PERCEIVE REMOTE SURROUNDINGS **152**
  ENERGY HEALING: HEALING A PERSON'S ETHERIC BODY, WHICH IN
    TURN HEALS THEIR PHYSICAL BODY **152**
    JESUS'S DIRECT AND REMOTE HEALING...**152**
    PAUL'S HEALING..**153**
    THE APOSTLES' HEALING **153**
  PSYCHOKINESIS: THE MANIPULATION OR MANIFESTATION OF PHYSICAL OBJECTS **153**
    JESUS'S COMMAND OVER FORCES OF NATURE AND THE MANIPULATION
     OF MATERIAL OBJECTS, SUCH AS FOOD...**153**
    BODILY PROTECTION FROM FIRE **153**
    BODY LEVITATION **154**
    SPIRITUAL MANIFESTATIONS INVOLVING PHYSICAL OBJECTS **154**
  NON-LOCAL CONSCIOUSNESS: ACCESSING SPIRITUAL ABILITIES
    OR TRAVELING OUTSIDE THE PHYSICAL BODY.....................................**155**
    OUT-OF-BODY (OBE) OR NEAR-DEATH (NDE) EXPERIENCES.....................**155**
    HUMANS ACCESSING HIGHER DIMENSIONS **155**
    INSTANCES OF DEAD HUMANS BEING BROUGHT BACK TO LIFE...................**156**
    NON-PHYSICAL BEINGS APPEARING TO HUMANS....................................**156**

BIBLIOGRAPHY 157

INDEX.. 164

ENDNOTES... 167

REFERENCE QR CODES.. 175

TABLE OF FIGURES

FIGURE 1: THE HYDROGEN ATOM, AS IT APPEARS IN OCCULT CHEMISTRY..22

FIGURE 2: A TORUS..22

FIGURE 3: THE HUMAN CHAKRAS...23

FIGURE 4: WALTER RUSSELL'S DEPICTION OF AN ATOM IN THE UNIVERSAL ONE..............................29

FIGURE 5: MATTER'S SPHERICAL STANDING WAVE STRUCTURE,
FROM SCHRÖDINGER'S UNIVERSE..29

FIGURE 6: DEPICTION OF AN ATOM FROM THE PRINCIPLES OF LIGHT AND COLOR..........................30

FIGURE 7: THE ANU AS ILLUSTRATED IN OCCULT CHEMISTRY..30

FIGURE 8: COMMON NAMES FOR EARTH'S WAVE-BASED DIMENSIONS. 40

FIGURE 9: A GEOMETRIC THOUGHT EXPERIMENT 43

FIGURE 10: SEPARATION OF A SIMULATED UNIVERSE – COSMIC CELL DIVISION (MITOSIS)? 45

FIGURE 11: THE ANU MOVES ENERGY BETWEEN THE ETHERIC AND ASTRAL DIMENSIONS...........46

FIGURE 12: MAGNETIC FIELD LINES 50

FIGURE 13: THE SHAPE OF AN IRON (FE) ATOM, AS DRAWN IN THE BOOK OCCULT CHEMISTRY. 50

FIGURE 14: DIMENSIONS HAVE VARYING DENSITIES..53

FIGURE 15: AN ELECTRO-MAGNETICALLY SHRUNKEN US SILVER DOLLAR 59

FIGURE 16: THE DIFFERENCE BETWEEN PROPHETS AND MEDIUMS 61

FIGURE 17: THE TESSERACT AND INFINITY 65

FIGURE 18: THE COSMIC OUROBOROS AS REPRESENTATIVE OF A
FOURTH-DIMENSIONAL INFINITY LOOP 66

FIGURE 19: SMALL TORI IN THE CORE OF A 3D-PRINTED TORUS 68

FIGURE 20: MATTER EXISTS WITHIN THE CORE OF TORI AT MULTIPLE LEVELS OF SCALE................68

FIGURE 21: A 3D SPHERE PASSING THROUGH A TWO-DIMENSIONAL
WORLD WOULD LOOK LIKE A GROWING AND SHRINKING CIRCLE 69

FIGURE 22: A 4D SPHERE PASSING THROUGH A THREE-DIMENSIONAL
WORLD WOULD LOOK LIKE A GROWING AND SHRINKING SPHERE..................................70

FIGURE 23: PROGRESSION OF BIRTH, GROWTH, MATURITY AND DEATH..70

FIGURE 24: THE FRAMEWORK I DEVELOPED FOR MY SPIRITUAL JOURNEY.....................................77

FIGURE 25: THE PROCESS OF SPIRITUAL EVOLUTION..83

FIGURE 26: CHART SHOWING HOW THE NON-PHYSICAL WORLD HELPS
THOUGHTS BECOME REAL..86

FIGURE 27: PARANORMAL ABILITIES ARE HOW SPIRITUAL BEINGS TRAVEL,
COMMUNICATE AND MANIPULATE OBJECTS...88

FIGURE 28: THE FIVE CATEGORIES OF PHYSICAL, MENTAL AND SPIRITUAL HEALING 91

FIGURE 29: A SINGLE GLOBAL WORLDVIEW CAN BRING TOGETHER SCIENCE, RELIGION AND METAPHYSICS..97

FIGURE 30: THE FORMATION OF EARTH'S NON-PHYSICAL DIMENSIONS 101

FIGURE 31: HOW DIMENSIONS MAINTAIN AN ENERGY BALANCE 103

FIGURE 32: THE PROCESS OF BIRTH AND DEATH FOR AN ETERNAL CONSCIOUSNESS..................104

FIGURE 33: SPIRITUAL ABILITIES BY DIMENSION AND NON-PHYSICAL BODY................................107

FIGURE 34: A CAUSAL ENTITY 113

FIGURE 35: MILO WOLFF'S WAVE-BASED THEORY OF EVERYTHING, REPRODUCED FROM SCHRÖDINGER'S UNIVERSE..120

FIGURE 36: THE BENDING OF LIGHT IN EINSTEIN'S THEORY OF RELATIVITY 121

FIGURE 37: A GEOCENTRIC (PTOLEMAIC) MODEL OF THE SOLAR SYSTEM, WITH THE EARTH AT ITS CENTER..124

FIGURE 38: OCCULT CHEMISTRY ILLUSTRATION SHOWING HOW ANU LINE UP WHEN EXPOSED TO ELECTRICITY..125

FIGURE 39: NEON ATOMIC NUCLEUS AND ELECTRON ORBITAL SHELLS 131

FIGURE 40: THE DIPOLES AND QUADRUPOLES THAT CREATE S- AND P-SHELL ELECTRON ORBITAL CLOUDS..132

FIGURE 41: THE QUADRUPOLES THAT CREATE D- AND F-SHELL ELECTRON ORBITAL CLOUDS 133

FIGURE 42: SEPARATION OF UP- AND DOWN-QUARKS IN THE HYDROGEN ATOM TO CREATE QUADRUPOLES 135

FIGURE 43: HYDROGEN ATOM WITH THE ADDITION OF QUARK-BASED ELECTRON ORBITAL CLOUD..135

FIGURE 44: MOVING THE OVOID-SHAPED OUTER SHELL OUTWARD, SO THE ORBITAL CLOUD CAN REMAIN INSIDE IT..135

FIGURE 45: CHANGING THE OUTER SHELL TO A CIRCULAR SHAPE..135

FIGURE 46: THE FIRST IMAGE OF A HYDROGEN ATOM 135

FIGURE 47: A 3D-PRINTED PHYSICAL REPRESENTATION OF A DONUT-SHAPED TORUS 136

FIGURE 48: SMALL TORI IN THE CORE OF A 3D-PRINTED TORUS 136

FIGURE 49: MATTER EXISTS WITHIN THE CORE OF TORI AT MULTIPLE LEVELS OF SCALE 137

FIGURE 50: TWO TORI DEFINED WITHIN THE EARTH'S MAGNETOSPHERE 138

FIGURE 51: TWO SHAPES OF A 3D PRINTED TORUS 138

FIGURE 52: LOOPING TWO TORI TOGETHER..139

FIGURE 53: TWO LOOPED TORI—THE FINAL SHAPE 139

FIGURE 54: A SPIRITUAL DEVELOPMENT FRAMEWORK..141

FIGURE 55: STEP ONE: A SPIRITUAL PATH ON-RAMP 143

FIGURE 56: STEP TWO: DEVELOP AN INTELLECTUAL UNDERSTANDING OF FAITH 144

FIGURE 57: STEP THREE: RECONCILE YOUR FAITH WITH YOUR LIFE EXPERIENCES 144

FIGURE 58: STEP FOUR: MATURE A UNIQUE SPIRITUALITY..145

FIGURE 59: THE ROLE OF JUXTAPOSITIONS IN SPIRITUAL DEVELOPMENT....................................146

FIGURE 60: MORAL COURAGE - THE INTEGRATION OF FAITH, HOPE AND LOVE FOR OTHERS.........148

FIGURE 61: THE "Z" AXIS..148

INTRODUCTION

A Word About Truth

Many times, people disagree on what's true because they don't understand the difference between subjective and objective truth, and it's easy to confuse one with the other. Just because a truth is based on a person's own (subjective) experience does not make it an "objective truth," that is, a truth that has been demonstrated through rigorous testing. Spiritual truths are often thought of as subjective truths based on a person's interpretation of a spiritual experience, which could vary with their level of spiritual maturity. Scientists, on the other hand, are trained to carefully test something before considering it true. In fact, scientists begin to consider that something might be true only after they have *repeatedly failed to prove it false!* That's why I have presented ways to dispute claims made in this book. It is my attempt to transform subjective views about the spiritual world into views that can be objectively tested and demonstrated—even if it's proven wrong.

This book is unique. Unlike other books about the spirit world that either are difficult to understand or merely scratch at the surface of the subject, *Earth's Hidden Reality* describes humanity's spiritual existence in an organized fashion and offers observations and suggestions for experiments to test its claims. Its conclusions are based on wave-based physics that, less than a century ago, was considered an alternate theory to quantum mechanics.[1] Through the application of wave-based physics, this book shows how the spirit world is as real as the everyday physical world in which we live and explains how it co-exists in "our" space.

Teaching about the reality of the spirit world typically has been the province of religious leaders, but they have been reluctant to discuss it in recent years. Similarly, scientists refrain from conducting research in this area, and spiritual experiencers, people who have had direct personal interactions with the spiritual, have often been called charlatans. Furthermore, the entertainment industry has sensationalized the idea of a spirit world, exaggerating aspects of it to the point where the possibility of having thoughtful conversations about its reality has been greatly diminished.

It's a sad state of affairs, but even with so much misinformation stacked against the acknowledgment of the spirit world's existence, it is becoming more and more clear that this hidden reality does, indeed, exist and will become a serious subject for broad scientific inquiry as solid, science-based experimentation supporting the probability of its presence is presented.

> **"Our materialistic cultural paradigm does not support our spiritual awakening. In fact it ridicules it and pathologizes it. But the only thing that can stand up to and transform the dominant scientific paradigm, as well as the dogmas of organized religions, is direct spiritual experience. Spirituality is something that is experiential; it reflects our direct experience of normally hidden dimensions of reality."**
> —Stanislav Grof, MD, psychiatrist and a principal developer of transpersonal psychology [2]

The reason I am so confident the spirit world exists is because I am a "spiritual experiencer" with nearly two decades under my belt of feeling, hearing, and seeing non-physical world inhabitants. It is an experience I did not seek (and at one time it frightened me), but now it is something I consider a gift. Between 2016 and 2019, I published three books on the subject, each attempting to not only relate my spiritual experiences but also to explain and clarify this hidden reality's infrastructures and processes. This fourth book is no different and contains updates based on my continuing experiences and investigations. Happily, many of the ideas discussed in my first three books have matured, allowing me to provide readers of this book with more complete explanations in places where there once existed only partially-developed thoughts.

This book is divided into five sections, which include Essential Concepts, Three Struggles and Appendices.

Essential Concepts

This section introduces you to terminology referenced throughout this book. You will learn more about these things as you read, but it is my hope that these quick summaries will provide a framework to help jump-start your understanding:

1 **The Theosophical Society**—Theosophy means "Divine Wisdom." The Theosophical Society is a world-wide organization focused on the study of divine wisdom and has existed for over a century. A person can belong to an organized religion and also be a theosophist. They are complementary.

2 **Anu and *Occult Chemistry*** [3]—Theosophical literature identifies the anu as the smallest item in physical existence. It spins and pulses inside an atom's nucleus and is responsible for its spin. This information was first published in the book *Occult Chemistry* in the early 1900s. The book contains over two hundred drawings and diagrams of atomic elements. While the information it presents was obtained in an unorthodox manner, science has independently verified things it discusses.

3 **Torus**—A torus is a donut-shaped electromagnetic field that can be found throughout the universe. Tori, the plural of torus, exist at multiple orders of scale in the universe and wrap around not only the human body (the heart torus) but the entire earth (the magnetosphere), the sun (the heliosphere) and even galaxies. Anu have a torus shape.

4 **Chakras**—The names and locations of human chakras, or energy centers, are widely known in the metaphysical community but not necessarily outside of it. The seven chakras are respectively tied to each of the seven levels of the astral

dimension. I cover the astral and earth's other non-physical dimensions in chapter three.

5 Quantum Entanglement—This is a Quantum Mechanics term that is discussed throughout the book. Science has been trying to understand this phenomenon for decades, but it has been used in metaphysical circles for centuries.

Three Struggles

For some readers, the concepts described in this book will be so beyond their immediate comprehension that they will reject them out of hand. Their rejection, plus humanity's inability to accept the existence of the non-physical world (what I call the spirit world) and its inhabitants is woven into three struggles I have identified as part of humanity's journey to and through earth's hidden reality. These struggles are to:

- **Discover It:** Believing the Non-Physical World Exists
- **Explore It:** Understanding Why the Non-Physical World Exists
- **Embrace It:** Making This New Reality Your Own

Struggle I — Discover It
Believing the Non-Physical World Exists

One reason why the average person simply can't imagine that the spirit world exists is because they rarely see objective, thoughtful discussions of the subject presented on television or written in print. It's much easier, it seems, to ignore the whole thing—or brush the spirit world off as entertainment at best or pseudoscience at worst. Especially since this invisible world is so … *invisible!*

Churchgoers also rarely hear the reality of the spirit world preached from pulpits or mentioned in Sunday school classes, places where one would expect to hear that its existence is more than a mere tenet of faith. Is it any wonder that my Christian brothers and sisters are just as skeptical of this reality as most scientists!? Perhaps it's because people who believe "in faith" these days are more likely than not to find themselves—and their beliefs—objects of derision if not outright scorn in the media and in movies. This kind of treatment makes them hesitant to consider the possibility that a reality beyond their own exists—even if it's likely to be true.

I was once in that place myself, ignorant of another world beyond my easily apprehended physical one—that is, until that other world began knocking on my door! In truth, it began knocking *inside my head*, an experience described later in this section. It is the most detailed (and most personal) section of the book, with stories gleaned from my many years of experiencing, questioning and learning about the non-physical world. Many of its further readings have their roots in my earlier books and have been expanded, updated and repackaged into bite-sized concepts here. I use them to make a big point about the non-physical world toward the end of Struggle One, so please take time to become familiar with each of these end-of-chapter ideas before moving forward. The topics include:

- Wave-Based Matter in History
- Does Angular Momentum Create Mass?
- Quantum Disentanglement on a Cosmic Scale
- How Different Dimensions Can Have the Same Vibrational Frequencies within Them
- A Wave-Based Explanation of the Double-Slit Experiment
- Prophets, Mediums and the Ethical Use of Spiritual Abilities
- Death as a Fourth Dimensional Process

A point I want to emphasize in chapters four through six is that the laws of physics remain unchanged across time—from the past, to the present and into the future. This includes scientific anomalies, miracles in the Bible and paranormal abilities. These chapters show how the estrangements between science, religion and metaphysics can be explained through wave-based physics and reunited:

Science Paradoxes (discussed in chapter four)
- A Theory of Everything (ToE)
- Strong and Weak Atomic Forces
- Magnetism
- Dark Matter and Dark Energy

Miracles in the Bible (discussed in chapter five)
- Telepathy
- Remote Perception
- Energy Healing
- Psychokinesis
- Non-local Consciousness

Paranormal Abilities (discussed in chapter six)
- UFO Appearance, Speed and Maneuverability
- Unseen Assistance—Invisible Helpers
- Mediumship and Channeling

Struggle II — Explore It
Understanding Why the Non-Physical World Exists

This section suggests steps to increase the probability of your having a spiritual experience. A kindly word of warning here: this struggle may cause you to question some of your religious beliefs. It has been my experience, though, that such spiritual reexaminations will enhance rather than undermine one's foundational spiritual beliefs and will ultimately produce a stronger spiritual understanding.

It is during this period of questioning that having a spiritual development framework can help provide perspective and clarity. I introduce the framework that I developed in chapter nine and review it in greater detail in appendix D. It's important to find one that you like, a process that will help measure your spiritual journey and keep you on-track.

The further reading concepts in this section are more complex and build on what we learned in Struggle One:
- My Spiritual Development Framework—An Overview
- Spiritual Emergencies—Where to Get Assistance
- The Power of Your Thoughts
- Healing Overview

Struggle III — Embrace It
Making This New Reality Your Own

There are more things to learn about the non-physical world than you can explore in your lifetime. This is why the chapters about embracing the spiritual are brief but convey big ideas—to encourage you to have big dreams of your own! Acting on your non-physical world discoveries helps create greater acceptance for earth's hidden realities, especially when they have the power to benefit humanity. The non-physical world is new territory, so take advantage of your early entry into it!

Chapter fourteen's concept is a big one, which is why I saved it for last. It's a conjecture about how a wave-based, multi-dimensional view of earth explains the creation of dimensions beyond the physical world, how energies remain balanced across those dimensions, and how hidden realities play into the process of birth and death. Researching and learning about the processes described herein also will provide a framework to help you better understand how earth's multi-dimensional environment functions.

Appendices

There are five appendices, which provide more detail on specific topics, some of which first appeared in my earlier books. They are:

- Appendix A—Observations and Experiments to Falsify Claims Made in this Book
- Appendix B—Wave-Based Atomic Forces
- Appendix C—A Hierarchical Torus Structure of the Universe
- Appendix D—Framework for Spiritual Development
- Appendix E—Wave-Based Explanations for Miracles in Bible Scripture

I'm the happiest with appendix A. It has been my goal for a number of years to provide a way to falsify (another word for "disprove") the claims that I make in this book. Just to set expectations, I am providing ideas for experiments, not protocols and methodologies, but I am proud of the fact that they're there. Skeptics have for decades asked people who believe in the spiritual (and in faith) to "show me evidence," and while it may not be what some people may call acceptable, it is a good start. The other appendices similarly dive into details, which provide information not covered elsewhere in the book.

What This Book is Not About

This book contains a science-oriented explanation of what can be described as anomalies, miracles, paranormal or supernatural. As its back cover states, "It's not magic. It's not miraculous. It's SCIENCE." This book is not about theology, religion or religious thought, but it does attempt to show how some things that people believe in faith can be explained through science as existing. Furthermore, this book is not about the paranormal, supernatural or metaphysical, but it does attempt to show how some techniques and abilities that these different groups teach can be explained through science as being real.

Consider the Corpus

Of both science-based skeptics and religious believers, I ask for patience as I present my case for the existence of the non-physical world. I also ask that you consider the corpus, the totality of what is presented. It's easy to dismiss a single idea, but a single "rejected" idea could one day become an important thread in the fabric we are weaving—part of a whole that's incomplete (or even incomprehensible) without it. So, again, I counsel patience as you struggle with unfamiliar, perhaps even threatening, concepts. This is precisely why there are so many ideas and examples presented in this book. I want the concept of a wave-based reality to be difficult to dismiss out of hand, because this is something I have struggled with as well—sometimes mightily! Having said this, I hope you enjoy the thoughts and questions that will arise as you read this book, as well as the lively conversations that may ensue from your discussions with others about so challenging a subject.

Now let's dive in.

ESSENTIAL CONCEPTS

These concepts are discussed throughout the book, so I introduce them here.

The Theosophical Society

Full disclosure: I am a Theosophist. This is what I decided after years of reading, writing and speaking about the non-physical world. I have found that Theosophical literature contains the most impressive, unbiased body of knowledge about the non-physical world.

The Theosophical Society was started in 1875 in New York City during a time of spiritual revival in America and has since grown into a worldwide organization. Today, its international headquarters is located in Adyar, Chennai, India, and its American section is headquartered in Wheaton, Illinois. It is dedicated to "promoting the unity of humanity; fostering religious and racial understanding by encouraging the study of religion, philosophy and science; and furthering the discovery of the spiritual aspect of life and of human beings." (from the www.theosophical.org web site)

The Theosophical literature I reference were written by Arthur Powell, Charles Leadbeater and Annie Besant—three of the Society's most prolific early authors. Much of the information they present was gathered through the use of spiritual abilities, but as Charles Leadbeater mentions in the introduction of one book,[4] steps were taken to ensure its accuracy:

- All information had to be confirmed by the testimony of at least two independently-trained investigators.
- Information was verified as true by either the author or by students of Theosophy with specialized knowledge when the author was not qualified to do so.

Arthur Powell's works are particularly good because each of his books source information from over two dozen other Theosophical books, which he references in the margin of each page!

Anu and *Occult Chemistry*

Theosophical literature characterize anu as the "ultimate atom," the smallest structure that humans on earth can observe. Figure 1 is an illustration of the Hydrogen atom taken from the book *Occult Chemistry*, which first appeared in 1908. The authors, Annie Besant and Charles Leadbeater, observed a Hydrogen atom containing a single proton and neutron, using unique visualization techniques. Within each proton and neutron, they noticed three round objects that seemed to be connected to one another. They didn't have a name for

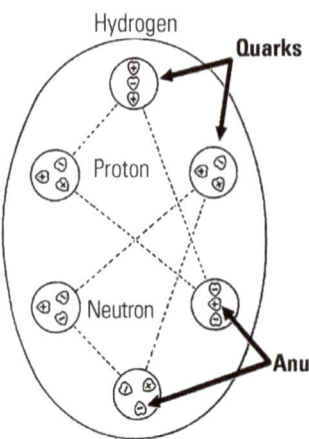

Figure 1: The Hydrogen atom, as it appears in *Occult Chemistry*

these objects back then, but we know today that what they described were up- and down-quarks, which were not discovered by scientists until 1964![5] The lines that connect the quarks are what we now call Gluons—also published in the 1908 edition of *Occult Chemistry.*

Torus

The torus (Figure 2), quite simply, is a donut-shaped geometric energy form. Its uniqueness comes from its complex spin. A torus can spin in three ways. The first spin is circular, the way an automobile tire spins. The second spin is an inside-out spin, in which

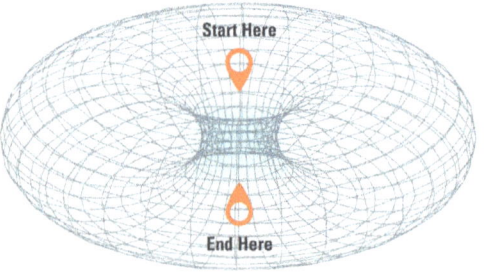

Figure 2: A torus

a point on the inside of the torus spins over the top of the torus and then underneath the donut shape back to the inner point where it started. The third type of spin is a combination of the circular spin and inside-out spin. Imagine sliding a pole across the middle of the torus, as if you were going to measure its diameter. Then, as the torus rotates around its pole, it also spins end-over-end.

Chakras

The seven primary human chakras (Figure 3), or energy centers in the body, correspond to the seven levels within the astral dimension, one of a number of spiritual dimensions that interpenetrate each other and exist beyond our immediate comprehension. They are often identified by color, with the lower chakras in the astral dimension having a warmer color (i.e., red, orange and yellow) and the higher chakras having a cooler color (i.e., blue, indigo and violet). People also associate the color(s) of a person's aura, the energy field that surrounds their body, with the different colors of the chakras—and, so too, with the different levels of the astral, with astral level 1 being lowest. I discuss later in the book how these can be related. The names are:

- **Root** (red)—The lowest chakra, located between the legs—astral level 1.
- **Sacral** (orange)—Located just below the navel—astral level 2.
- **Solar Plexus** (yellow)—Located above the navel and below the sternum—astral level 3.
- **Heart** (green)—Located at the heart - astral level 4.
- **Throat** (blue)—Located at the throat—astral level 5.
- **Third Eye** (indigo)—Located above the nose, between the eyebrows—astral level 6.
- **Crown** (violet or clear/crystalline) —Located on top of the head—astral level 7.

7 Crown
6 Third Eye
5 Throat
4 Heart
3 Solar Plexus
2 Sacral
1 Root

Figure 3: The human chakras

Quantum Entanglement

Quantum entanglement is a Quantum Mechanics phenomenon that occurs when pairs or groups of particles become inter-connected in such a way that a change in one particle immediately occurs in all the others, regardless of the distance in which they are separated—even across a galaxy![6]

Scientists have been trying to understand this phenomenon for years, but it has been regularly used for centuries in metaphysical circles to allow energy healers to remotely attach to, and heal, a person's wave-based body, among other things.

STRUGGLE I —
DISCOVER IT

Discovery is an adventure—for an explorer sailing the seven seas in search of a "new world" or for a spiritual adventurer like me (or perhaps you) seeking a better understanding of a world that lies just beyond our ability to perceive it.

This section recounts some of my spiritual adventures in that world over the past two decades and presents science-oriented concepts that help explain them. Taken together, these personal experiences navigating the non-physical, multi-dimensional world—and the findings that help explain that world—provide a wave-based, multi-dimensional explanation for the process of death. It shows how our existence is truly eternal and what we interpret as death is a process repeated in multiple dimensions. The section ends with a proposal (with justification) that our reality is composed of four spacial dimensions—not just three.

CHAPTER ONE
How My Spiritual Experiences Started

I began having spiritual experiences in 2003. It wasn't my intent to have them, but they have persisted—and have never gone away. My first encounter was with a spirit that entered my head and starting moving around my brain—at least that's what it felt like! It was this encounter (and many others like it) that inspired my research into, and writing about, the non-physical world and its inhabitants.

The early part of my spiritual journey began with learning how to feel the energy that flows through my body and experimenting with Christian deliverance ministry commands, another name for performing exorcisms. I began with a set of exercises to sensitize my body to the movement of energy within it and, after a while, reached a plateau where I could:

- pull energy into my body through an arm or a leg, move it through my body and then push it out through another limb,
- extend my arms and, while pressing my palms together, move energy in a loop through my arms and upper chest, and
- draw energy in through the crown chakra in the top of my head, move it down into my chest, and loop the energy multiple times up and down the length of my torso.

My interest in sensitizing my body to the movement of energy within it came after a friend, who later became a mentor, pointed out to me that a sensation I felt in my hands came from my body energy. To see if you can sense energy coming from your body, perform this test:

1 First, stand up and vigorously shake your hands toward the floor as if trying to shake off a pair of gloves. Try to release as much tension from your shoulders and upper arms as you can. These movements are intended to make the feeling in your hands more obvious and intense.

2 Slightly curve the palms of both hands inward, cupping them into the form of a small bowl.

3 Turn the palm of your cupped left hand on top of the palm of your cupped right hand at a right angle, so that the fingers of your left-hand hover over the thumb of your right hand.

4 Hold your hands slightly apart—between 1–2 inches. Now, without letting them touch each other, slowly move your palms toward and away from each other, about an inch or so in each direction. Do this for one minute. If you can feel a slight pressure or resistance when you bring your hands together, similar to the pressure felt when you bring the like poles of two magnets together, then you are feeling what I described to my friend. He gave me a set of exercises to develop a sensitivity to my body's energy. It took about eighteen months of casual work on weekends to develop the skills necessary to control the movement of energy inside my body. If you choose to do this work, it's important to find an experienced practitioner who can teach you how to safely perform the exercises to develop your body's energy sensitivity. After I learned how to control the movement of energy within my body, I went on to the next step: purchasing a book about deliverance ministry.

Francis MacNutt's *Deliverance from Evil Spirits* taught me how to issue delivery commands. While lying in bed on a Saturday morning in May of 2003, I gave MacNutt's commands a try. Honestly, I was quite skeptical that mere words would elicit a response—but it worked! The sensation felt like multiple points

Confirming My Beliefs

After these experiences began occurring with some regularity, I started changing the delivery commands to either include or exclude the name of Jesus. When I realized that spirits would not move unless my commands were issued "in the name of Jesus," it convinced me that Jesus and the spirit world were real and drove my desire to learn more about it.

of energy moving outward from my spine to the side of my body. It usually took about five minutes before anything happened, but the spirits moved consistently each time I commanded them. After about four weeks, however, they became larger and were not as cooperative as the ones I had experienced earlier. These were the spirits that triggered my quest to master deliverance ministry techniques.

This is the story I mentioned at the beginning of this chapter; it was the first time a spirit took up residence in my head and decided to stay. It felt like a half-inch diameter sphere was moving around in my brain. When I commanded it to leave, it would move up to the inside of my skull but not pass through it. One morning as I was driving to work, I even commanded it to listen to music by Christian singer Amy Grant that I played in the car. This didn't work either, even when I turned the music up loud, although it did attract the attention of other drivers.

After trying unsuccessfully for a week to get my uninvited guest to leave, I returned to MacNutt's book for advice. In one chapter, he wrote that if you asked the spirit its name, its response could provide insight on how to get it to leave. So, I commanded it, "Tell me your name!" The response came back almost immediately. Its name was Jealousy. "Who was I jealous of?" Almost as soon as I expressed that wonder, the names of seven people rolled through my mind. Since I was more interested in getting rid of this spirit than debating whether I might be jealous of these people, I stopped and prayed, "Father, please forgive me for being jealous of so and so," and said each name aloud. The instant I spoke the seventh name, the spirit shrank to the size of a pea and drifted out of my head.

I was learning!

Further Reading – Wave-Based Matter in History

The wave-based theory of matter is foundational to the concepts put forth in this book. Wave-based thinking is not new; in fact, it competed with Quantum Mechanics up to the mid-twentieth century and enjoyed a number of influential supporters, including Erwin

Schrödinger, Ernst Mach and Albert Einstein.[7] Inventor Nikola Tesla similarly encouraged people to think in terms of "energy, frequency and vibration" when he spoke about the universe.[8]

Below are four references to wave-based matter that span over a century and are amazingly similar in their description. The first two references discuss the type of motion wave-based matter should possess. The last two references consider how that motion may come about at the atomic level:

1 Walter Russell was an artist and prolific twentieth-century spiritual writer who authored a number of books regarding spiritual reality. In his 1928 book, *The Universal One*, Russell drew his conception of an atom, which he claimed had been conveyed to him by spiritual beings[9] (Figure 4). It's hard to imagine what this structure might look like when viewed on the printed page, but I believe it is a top-down view of a three-dimensional torus, a spherical geometric shape with a shaft through its center.

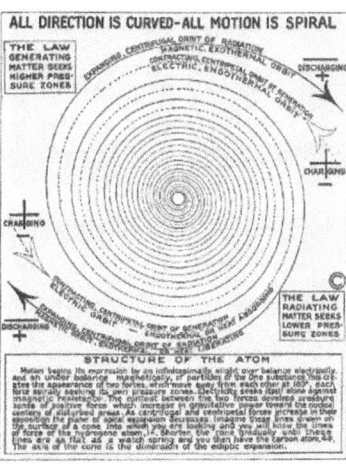

Figure 4: Walter Russell's depiction of an atom in *The Universal One*

2 Milo Wolff's 2008 book, *Schrödinger's Universe and the Origin of the Natural Laws*, depicts matter as a set of waves in space instead of a discrete object. Mirroring Walter Russell's comments, Wolff describes matter as a set of concentric waves (like those in a pond) that expand and contract, as shown in Figure 5.[10] Again, this

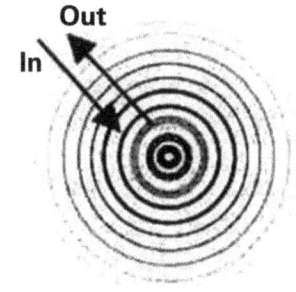

Figure 5: Matter's spherical standing wave structure, from *Schrödinger's Universe*

image could be a top-down view of a three-dimensional torus. Wolff and Russell both describe the atom's in-and-out motion, but notably in slightly different directions.

Russell's and Wolff's descriptions closely resemble the next two depictions of a wave-based atom.

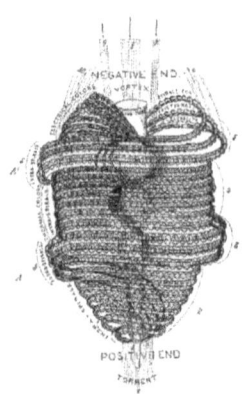

Figure 6: Depiction of an atom from *The Principles of Light and Color*

3 An atomic structure with wave-based characteristics (Figure 6) was described by physician Edwin Babbitt in his 1878 book *The Principles of Light and Color*. This image shows how this torus-shaped atom, an object smaller than today's subatomic particles, moves energy that passes through it, from one end to the other. He described the energy entering the atom as having a vortex shape, exiting it as a torrent. Babbitt spent the first part of his book's third chapter describing each component and theorizing how it could spin, expand and contract, based on heat.

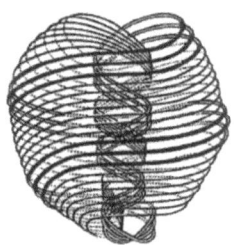

Figure 7: The anu as illustrated in *Occult Chemistry*

4 Three decades later, in 1908, Annie Besant and Charles W. Leadbeater, prominent figures in the Theosophical Society (its views were ascendant at that time), published illustrations of an atom very similar to Babbitt's in their jointly-authored first[11] edition of *Occult*[12] *Chemistry*. This was the anu, (Figure 7), a Sanskrit word that translates as "fine" or "atom." *Occult Chemistry* describes the anu's motion as a combination of spinning and vibrating. This is significant, because it matches the description of a fourth-dimensional (4D) spin called a Clifford Displacement[13] in which a 4D object rotates simultaneously in two separate dimensions. The presence of a 4D spin implies that part of the anu resides in the fourth

dimension, enabling it to serve as a bridge between dimensions. The anu's behavioral characteristics are central to the wave-based concepts discussed in this book, which—as far as its acceptance goes—might be compared to eating sushi. The idea of eating raw fish on rice may seem strange at first, but once you get over what it is, it tastes pretty good!

Anu and Strings

When an anu is taken apart (if you can imagine such a thing), it consists of ten separate loops of a string-like material. This aligns with a prediction of String Theory, that matter—in its smallest component—consists of tiny loops of strings.

CHAPTER TWO
The Door to the Spirit World Opens

The Early Years (2003-2007)

It took about six months of reading and practicing to figure out how to consistently remove spirits from my body using deliverance ministry techniques, and I became very good at it. Within minutes of a spirit entering my body, I was commanding it to leave, and if it didn't move fast enough (for me), I would manipulate my body energy to push it to a place where I could pull it out of my body with my hand. For example, if a spirit came into my head and wouldn't leave, I would use my body energy to push it against my skull, then grab its energy field with my hand and pull it out. It was strange at first but quickly became second nature and no longer strange.

I remember a few significant moments from this time, especially lying in bed one night when I caught a spirit in my hand as it was trying to enter my body. I held my arm straight up with the spirit inside my clenched fist. For about thirty seconds, this entity shook my hand trying to get out.

Deliverance as a Ministry?

In 2006, I became interested in offering deliverance ministry services as a way to help others. However, I quickly realized that deliverance ministry should be practiced only under the authority of an organization with extensive experience and should never be attempted by yourself. Spirits have varying abilities and strengths. If someone seeks you out for deliverance ministry, chances are the spirits in their body may be much stronger and more capable than you've experienced. Do not try to help them. Instead, refer them to someone who performs deliverance ministry work under the authority of a church or other organization.

It felt like I was holding onto a gyroscope! When I released it, the spirit hovered over my head for a couple minutes, which I could feel but not see, and my sense was that it couldn't believe what had just happened.

On two occasions, I was able to kick every spirit out of my body. This released a floodgate of joy, which was overwhelming. The net result of both experiences was that I learned how to distinguish "my" thoughts from those of other entities (i.e., spirits). This allowed me to know when thoughts were being placed into my mind. My rule of thumb now is to assume that any random thought or temptation that appears in my mind has been placed there by spirits. If these thoughts or temptations are embellished, however, I know it is MY mind taking over, and when that happens, I see it as a warning to stop and establish control over those thoughts.

Sometimes, I felt a spirit come into my head and within a few seconds began seeing images. They came in two forms, which I differentiate as dreams and visions. In a *dream*, if I opened my eyes while the images were playing, they would immediately disappear and I could see my physical surroundings. In a *vision*, however, the images remained superimposed ON TOP OF what I saw with my opened eyes. If what the Bible described as "visions" were anything like what I experienced, I can understand why people during those times took notice of them. I also had multiple experiences where a dream or vision consisted of an event during which I was presented with a choice. It made me think that spirits use dreams and visions to test us in situations that would be impractical to do while we are awake.

It took me about three years to become comfortable feeling spirits around me—mostly out of necessity because I could no longer maintain the tremendous mental energy required to expel spirits each time I sensed them in my body. It was only after accepting their presence that I realized that the vast majority of these spirits were harmless. Even those I thought malevolent were performing a beneficial service!

During this time of discovery, my reading interests continued changing, to more about religion and spirituality than business. I also had regular, though impromptu, meetings with my spiritual mentor during this period, who helped me understand what I was experiencing. Fortunately, he had trained for Christian ministry. This enabled him to frame my experiences in a religious perspective, which I greatly appreciated, because that was the only framework I had at that point for understanding my spiritual experiences.

Questioning My Faith (2008-2009)

These were years in which I seriously questioned my religious faith. If you look at the Spiritual Development Framework in appendix D, I was stuck at *The Wall*. For readers who are similarly questioning their religious faith, it will be helpful for you to know that I was able to move past *The Wall* only after I gave up my belief in the Trinity, because at that time (2009), I could not reconcile it or explain it. A few years ago, however, I gained a more spiritual understanding of the Trinity and have reincorporated it into my beliefs.

Building a Unique Spirituality (2010-2013)

After reconciling my beliefs, I was prepared to embrace a new spiritual worldview. This change was not without consequences, however, because friends and family were not always supportive of my new outlook. For instance, my wife had for a long time

Concentration

During this time, I was spending about an hour each day lying in a recliner trying to quiet my mind. Books I read on concentration recommended focusing on breathing to distract me from the activity in my mind, but I found it easier instead to focus on the inside of my closed eyelids. My reading on spiritual topics became more intense during this time and, at some point, I stopped ordering business and technology books (a watershed moment), which until then had been my favorite reads.

been concerned about the increasing changes in my life, and we divorced during this time. (Studies show that people who have spiritually transformative experiences also have higher incidences of significant life challenges, including divorce.[14]) This change was not pleasant but allowed me to begin building a new life. I started writing my first book during this time, in November of 2011. It's ironic, but I actually started writing about my experiences as therapy to help me through the pain of divorce. It was only later when I realized that I had the beginnings of a book.

Broadening My Understanding of Other Faiths (2014-2015)

One experience I very much enjoyed as I explored spiritual options was Saturday morning Torah study at a local Jewish synagogue. I had taken three classes on Judaism elsewhere, all taught from a Christian perspective. This included a class at my Manhattan church that lasted three months! I thought I knew what Judaism was about before attending the Torah study but quickly learned of my error as the teaching went on. This congregation was very spiritual, and I found myself in theological agreement with much of what they said.

First, the Jewish Tanakh (the Hebrew Bible, which Christians call the Old Testament) is written in Hebrew. While it may not sound significant, it is, because this common language serves as a bridge between ancient Jewish writings and modern Jewish culture, allowing today's Jews to reach across time, geography and circumstance to enrich their understanding of scripture and their history. Second, rabbinical scholars have studied The Torah, the first five books of the Hebrew Bible, for thousands of years and have amassed a treasure trove of allegories and cross-references for these passages. This was one of the reasons I enjoyed the synagogue's Torah study. I was constantly learning new things and gaining fresh perspective.

I highly recommend that you seek to understand a second faith tradition—in addition to the one you already know. It will enrich your spiritual development to see what areas of your faith are similar—and different—from others. Be sure to speak to senior faith leaders before attending a class or service, and always keep in mind that you are a guest in their place of worship. As my way of acknowledging this (and to set expectations), I told the faith leaders at the synagogue that I never would consciously say or do anything contrary to their teachings, and any comments I made would be meant to edify their congregation. I didn't compromise my beliefs; I was just careful about what I said. As a result, I never had a problem and made some good friends.

Published Works (2016-2019)

My first book was published in 2016 with its next two editions in 2017 and 2019.[15] The experiences described above (and more) combined to whet my curiosity for greater spiritual understanding, the fruit of which can be seen in this book. When I held the first copy of the first book in my hands, all I could think about was the challenging journey I had taken to get to that point.

Feedback on My First Book

One of the most memorable moments in my life as a newly minted author occurred in 2016 in Tucson, Arizona. I attended a science and consciousness conference there, so I could talk to attendees about my book. I had not previously attended the event, so I was completely unaware that most of the attendees either had their PhD in physics, neurology, etc.—or were working on it. One of the first people I introduced myself to was Dr. Stephen LaBerge, the professor who had proven that Lucid Dreaming was real. Since he reminded me of Emmet "Doc" Brown in the movie *Back to the Future,* I shouldn't have been surprised when—just like Doc Brown—LaBerge didn't immediately respond when I talked about my book. Instead, he just smiled, tilted his head and looked at me! It was then that I realized my book needed more science and less speculation. Thank you, Dr. LaBerge, for your very important, though wordless, feedback!

I first spoke about matter being wave-based in the second book and included thoughts and illustrations on the atomic structure of matter. What, you may ask, could I possibly have to say on the subject given that my background is in project management, not physics or quantum mechanics? Looking back now, after a great deal more study, I found that much of what I wrote in book number two was reasonably sound but somewhat vague. Many of the insights expressed in that book came to me as I built physical models of the nucleus of different atomic elements using 25mm magnetic hematite spheres. I figured that the magnetic attraction of the spheres would simulate the forces found in the actual atoms, and I am confident this was correct.

My third book focused more on the infrastructure of the non-physical world, and its diagrams heavily leveraged information gleaned from a number of Theosophical books. I had almost finished editing my third book when I discovered the information in these books, which led me to rewrite a number of chapters—these books are that good! The information in them is clear, and the diagrams are rich with information. Check listings for Besant, Leadbeater and Powell in this book's bibliography. Those highlighted with an asterisk "*" I believe are most important to read and understand.

Two weeks after my third book was published, a former manager called to ask if I would take a consulting position working for her. Since she was an exceptional manager, I took the offer and worked full-time for two years, during which I had time to think about and refine some ideas that appeared in my third book, and now they appear here. I was active on other fronts as well. For a number of years, I had been the organizer for a consciousness-related meetup-dot-com group. In 2020, after COVID hit, a friend who was running a similar group in a neighboring city and I began holding joint meetings via Zoom. Today, we have attendees dialed in from Canada, Mexico and the east and west coasts of the United States and have had presenters speak to us from Europe. Online meetings have enabled us to significantly expand our reach—and our community.

Further Reading – Does Angular Momentum Create Mass?

In a 2016 blog post, Edward Close, PhD, described a demonstration in his undergraduate physics course some years back in which the professor asked students to carry a suitcase-like wooden box from the back of the lecture hall to him up at the front.[16] Close's classmates found the box easy to carry straight to the front but difficult to turn when they tried to walk it over to the professor. The professor later explained that the box's resistance to turning was caused by the Gyroscopic Effect, an "angular momentum" he had created by mounting a rapidly spinning a flywheel inside the box. The box was easy to carry as long as it moved parallel to the direction in which the flywheel was spinning but was difficult to move perpendicular to the spinning flywheel because of the wheel's angular momentum. The best example of the Gyroscopic Effect, of course, is the spin of a gyroscope. As long as the flywheel is spinning, the gyroscope remains upright and will not fall over. A more familiar, and perhaps surprising, example is the wheels on a bicycle. Once the bike starts moving, the angular momentum created by the spinning wheels keeps the bicycle upright as it moves along.

Dr. Close used this classroom example to speculate that an object's mass[17] comes from the cumulative effect of the angular momentum generated by millions of spinning up- and down-quarks within an object's atoms.[18] It's an intriguing topic because resting mass in an atom—that is, the mass of the atom when it is not moving—only accounts for about 2% of its total mass. The rest comes from kinetic mass, moving mass that includes the action of those spinning quarks. This phenomenon is a necessary component to explaining a number of "other-worldly" anomalies.

What, for instance, will happen if quark spin on one or more of our reality's three-dimensional (i.e., height, depth and width) axes can be controlled? Imagine you have to move a 100-ton block of granite. If you could stop the spin (i.e., the angular momentum) of quarks in all three axes—all three dimensions—the block's mass (as a result of it having no angular momentum) would become almost zero, making it easy to lift and carry. Furthermore, if this massless block was placed between two other granite blocks such that there was a small gap between them, it may be possible to deform the block until it completely fills the gap. Do you like how everything looks? If so, you may now have the tools you need to build the Egyptian pyramids!

The idea of using levitation to build the pyramids is corroborated in Theosophical literature.[19] A century-old Hindu book similarly discusses how levitation is an ancient yogic technique that was once generally known and how Gautama Buddha, among others, made use of it.[20]

CHAPTER THREE
An Introduction to Wave-Based Dimensions

Figure 8 lists the non-physical (wave-based) dimensions by their most common names, although illustrating them in stairstep fashion, with one dimension pictured higher or lower than another, is wholly inadequate. They actually "interpenetrate" each other, meaning that they co-exist in the same physical space but at separate "frequencies," just as TV and radio waves co-exist in a physical space, each station on its own frequency. Note also that beings in each dimension have their own "body," listed on the right side of the diagram and separately drawn on the left side. Because of innate differences, which we will discuss later, it is impossible for a body from one dimension to exist in another dimension—although they can

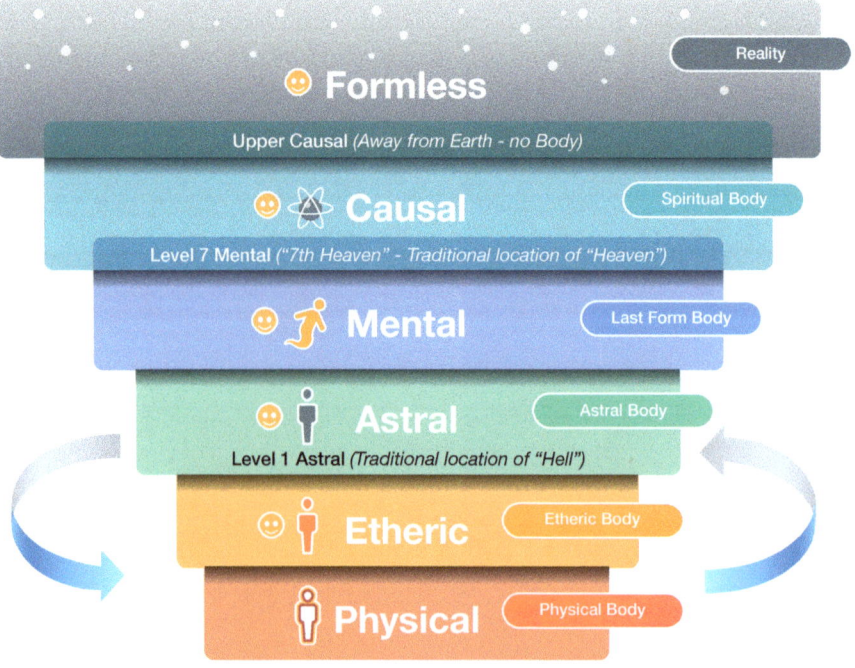

Figure 8: Common names for earth's wave-based dimensions.

interpenetrate each other. We'll discuss the different dimensional bodies later in this chapter.

A brief description of each dimension is presented below. The best sources I have found for this information are Theosophical books by Charles Leadbeater and Arthur Powell, which I reference in this chapter's endnotes.

The wave-based dimensions found on earth are the:

- **Physical**—The dimension we're most familiar with, the place we live and interact with one another. A wave-based consciousness can inhabit a (wave-based) human, animal or even a plant body here in "our" dimension.
- **Etheric**—A near mirror-image copy of the physical dimension, where objects and bodies interpenetrate one another, co-existing in the same space but just a slight frequency shift away.[21] This is possible because the etheric body is composed of energy and does not have a defined form. It works in conjunction with the physical body and supplies its energy needs. The etheric body, according to metaphysical literature, separates from our physical body during an Out-of-Body Experience (OBE). It is our consciousness in the etheric body that travels around the etheric dimension during an OBE, which is why it is never seen.
- **Astral** – Separate from the physical/etheric dimensions but also interpenetrated[22] with them. This is the dimension where our (wave-based) consciousness resides between its physical lives on earth, as noted by the two curved arrows in Figure 8. The astral dimension has seven levels,[23] as do the other dimensions. The lowest astral level is the traditional location of what is referred to as "hell" in religious literature.[24]
- **Mental** – This is the last dimension in which our wave-based consciousness inhabits a body with any form. In the lowest four levels, these bodies have a wispy form, but starting in the fifth level of the mental dimension, our consciousness inhabits a wholly spiritual body. The fifth through seventh levels of the mental dimension are often referred to as the "higher

heavens" in metaphysical literature.[25] The "seventh heaven," the seventh level of the mental dimension, is the heaven referred to in Jewish and Christian religious scripture.

- **Causal** – Beginning in this dimension, our consciousness has the ability to leave earth, just as a consciousness from another planet can come into the earth realm. Before moving into the causal dimension's third level, our consciousness sheds its spiritual body to become formless—a being without a body or a form.[26]

- **Formless** – Our "true existence" is as beings without form or shape. Very little is known about the formless dimension or dimensions that may exist beyond that. Once beings ascend into the formless dimension, according to Theosophical literature, they can no longer return to any of the form-based dimensions on earth.

How Wave-Based Realities Co-Exist in the Same Space

The concept of interpenetration is a fundamental part of wave-based physics because it explains how multiple realities can co-exist in the same physical space. See the sidebar for an analogy that explains this concept.

How Our Consciousness Uses Different Bodies to Move Between Dimensions

Two stumbling blocks that theologians face when they claim consciousness is eternal are the Laws of the Conservation of Mass and Energy. These laws state that in a closed system, such as our universe, mass and energy within it can neither be created nor destroyed. Skeptics

Interpenetration Analogy

Imagine a large, empty glass jar sitting on a table alongside small piles of stones, pebbles and sand. If you fill the jar with the stones, is it full? No, because you can still pour pebbles into the jar and fill the spaces between the stones. Is the jar full now? No, because you can still add sand to fill the spaces between the pebbles. When the jar is full, the pebbles and sand are said to interpenetrate the same space as the stones.

have used these laws to force theologians to acknowledge that because they didn't know where in the material universe energy from a human being's consciousness went after death, they could not make the claim that our consciousness is eternal. In a wave-based reality, this is not an issue because our consciousness can move inter-dimensionally—within a closed, multi-dimensional universe!

Consider the thought experiment shown in Figure 9. Imagine a cube, a square, a line and a point are floating in space in front of you. Notice that you can turn the cube until it looks like a square. You can do the same with the other shapes: turn the square to look like a line and the line to look like a point. But this process does not work in reverse. You can't turn a point to look like a line or a line to look like a square or a square to look like a cube. Another way of interpreting this thought experiment is to think of it in terms of objects within a dimension. Objects in higher dimensions can step down one dimension, but objects in a lower dimension cannot step up a dimension. Additionally, if a higher dimensional object wanted to step down two or more dimensions, it would first have to assume a shape in the dimension immediately below it so it could drop to the dimension below that. In essence, a higher dimension object has to inhabit objects in each lower dimension

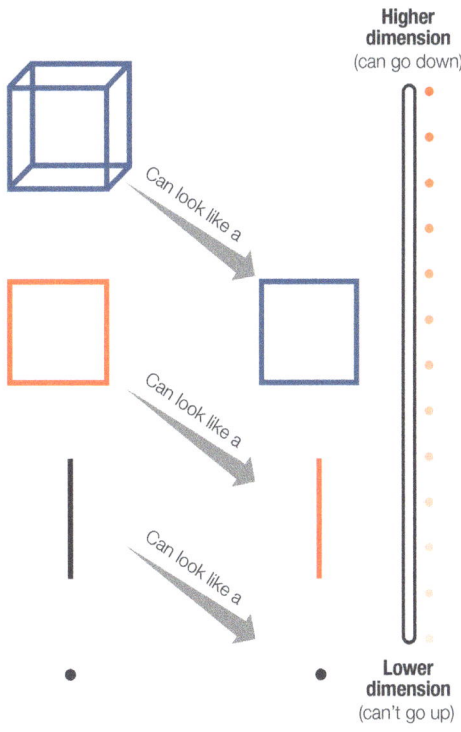

Figure 9: A geometric thought experiment

have to assume a shape in the dimension immediately below it so it could drop to the dimension below that. In essence, a higher dimension object has to inhabit objects in each lower dimension in order for it to descend through more than one dimension.

Now let's interpret this thought experiment in terms of consciousness. If a consciousness in the fifth level of the mental dimension[27] is to become embodied in the physical dimension, the process I just described suggests that it would have to drop down to a lower-level of

Interpenetration and Many Worlds

The Many Worlds Interpretation of Quantum Mechanics suggests that many worlds exist in parallel in the same space and time as our own. This can occur in fully wave-based physics, based on interpenetration. Many worlds can co-exist in the same physical space by vibrating at different frequencies.

that dimension to acquire a mental form body. It would then use its lower mental dimension body to descend to the astral dimension, where it would acquire an astral body. It would then use its astral body to drop down into the etheric dimension, where it would acquire an etheric body. The consciousness in its etheric body, with the other bodies wrapped around it on the inside, is then joined to the body of a fetus (i.e., a baby in the womb) in the physical dimension. Metaphysical literature corroborates this, saying that human beings have an associated body in each dimension. Some speculate that these higher-dimensional bodies are what form the human aura.

Further Reading – Quantum Disentanglement on a Cosmic Scale

A 2015 *Nature* magazine article suggested that the physical universe could be held together by the quantum entanglement of the objects within it.[28] Quantum entanglement is a phenomenon of Quantum Mechanics that occurs when pairs or groups of particles become interconnected in such a way that a change made to one particle immediately occurs in all the others, regardless of the distance separating them.[29]

The article's author constructed a computer model in which he gradually reduced the quantum entanglement of objects on the balloon-like outer surface of a simulated universe (Figure 10), to see what would happen when quantum entanglement was reduced to zero. He called this act "quantum disentanglement!"

When the computer simulation completed, the author discovered that when quantum entanglement was reduced to zero, **the model universe split in two!** This is an important finding from a wave-based physics perspective, because it could explain a scientific paradox—what happened to anti-matter, matter with a charge opposed to that of matter, which is predicted to have been created along with matter during the Big Bang. Equal parts of both matter and anti-matter were formed when the universe was created, but today, very little anti-matter can be found within it. Where did the anti-matter go? Given the simulation just described and our knowledge of wave-based dimensions, one could theorize that anti-matter might have become the "matter" of another, completely separate, dimension! Theosophical literature does state that numbers in the astral dimension must be read backwards, which would support this conjecture.[30]

Could quantum disentanglement also explain how the non-physical dimensions around the earth were created? This idea

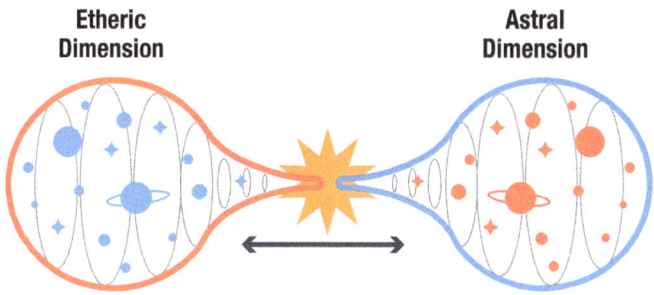

Etheric Dimension   **Astral Dimension**

Figure 10: Separation of a simulated universe –
cosmic cell division (mitosis)?

reminds me of the biological process of mitosis, or cell division, but on a planetary scale! Could earth's creation be explained as being the result of multi-dimensional mitosis?

Could Anu Be the Source of Cosmic Motion…

The inter-dimensional clash between matter and anti-matter presents opportunities for other ideas as well. It is well known that when matter and anti-matter meet, they annihilate each other in a violent explosion! If you are a fan of the TV series *Star Trek*, you may recall that the starship Enterprise was powered by the reaction of matter with anti-matter. In the further reading at the end of chapter 1, we discussed a characteristic of the anu described in *Occult Chemistry*—that it both spins and pulses. Since this three-dimensional object exhibited a fourth-dimensional spin, we concluded, as did the

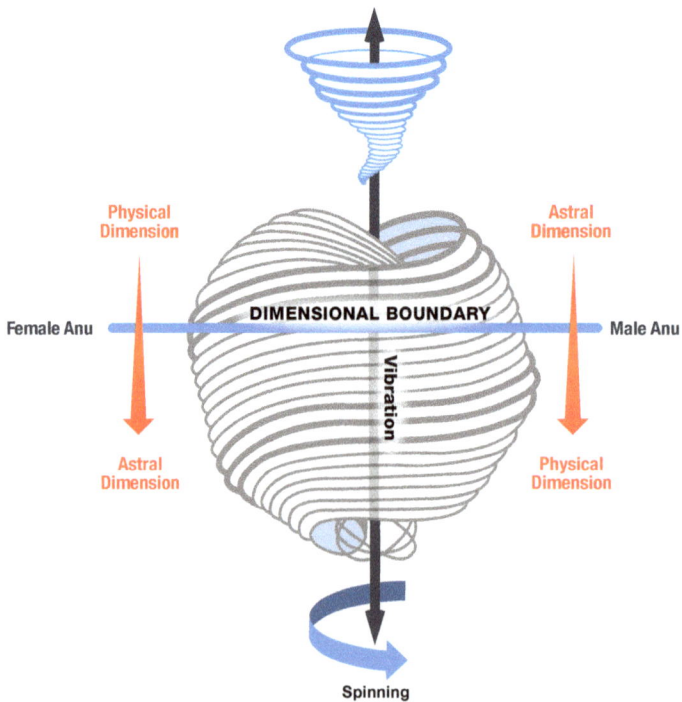

Figure 11: The anu moves energy between the etheric and astral dimensions

authors of *Occult Chemistry*, that anu move energy between the astral and the physical (actually the etheric) dimensions.[31]

What keeps atoms, planets and galaxies spinning and moving through the physical universe? Could it be possible that the source of the movement of every object in every dimension is the anu, with its power coming from the interaction of matter with anti-matter as it moves between the astral and etheric dimensions (Figure 11). While it sounds outlandish, this insight could make sense if anti-matter is matter in the astral dimension. Metaphysical texts say the astral dimension is form-based. Could the movement of objects in the physical universe be the result of minute matter/anti-matter explosions occurring inside anu?

...and the Colors in the Human Aura?

The presence (or not) of an aura around human beings continues to be debated. A number of books and videos have been produced that argue both for and against its existence. Two references I like, which support the existence of an aura, are Walter Kilner's *The Human Atmosphere* and Charles Leadbeater's *Man Visible and Invisible.*

Dr. Walter Kilner used chemical-stained glass screens to physically view and document the aura of over forty patients at a hospital in London in the early 1900s. In his book, he expressly states that he was able to view the aura by physical means only, and that he did not use any occult method to do it.

In contrast, Charles Leadbeater, an early leader of the Theosophical Society, used clairvoyance to view people's auras. What makes his book about auras, titled *Man Visible and Invisible*, interesting is its chart listing the different colors found in an aura and what they each mean. His book also contains a number of illustrations that interpret different patterns and shapes that can be seen in an aura.

Regarding auras and anu, *Occult Chemistry* mentions that anu light up with seven different colors, corresponding to the seven levels in the astral dimension.[32] If an anu can pull energy from any of the seven levels of the astral dimension, one might conclude that the colors emanating from anu are what create the different colors seen in the human aura.

> **"The world is not prepared yet to understand the (Hidden) Sciences ... that there are beings in an invisible world ... and that there are hidden powers in man which are capable of making a God of him on earth."**
> —Helena P. Blavatsky, co-founder of the Theosophical Society

CHAPTER FOUR
Explaining Science Anomalies and Paradoxes with Wave-Based Physics

Wave-based physics can explain science anomalies and paradoxes, miracles found in scripture and paranormal abilities, as the next three chapters will demonstrate. In this chapter, I explain how wave-based physics can explain scientific anomalies and paradoxes, which include:

A Theory of Everything (ToE)

Particle physicists have struggled for decades to develop a "Theory of Everything," a set of formulae that together explain all physical phenomena, from the smallest particles to the largest celestial objects. Wave-based proponent Milo Wolff, PhD, published a wave-based ToE in his 2008 book *Schrödinger's Universe*, which is summarized in appendix A. Dr. Wolff's book contains a number of mathematical and scientific justifications for wave-based physics in addition to history and commentary.

Strong and Weak Atomic Forces

Both strong and weak atomic forces may originate from the spin of three anu within each quark.[33] Anu move energy between the astral and the etheric dimensions and are characterized as being either positive or negative, depending on the direction in which they move energy between the astral and etheric dimensions.[34] Anu that move energy into the etheric dimension are said to be positive and anu that move energy into the astral dimension are said to be negative.

The presence of an odd number of anu (three) in each quark results in an imbalance of astral energy either entering or leaving the quark. As a result, a positively charged up-quark (two + anu to one – anu) will have more astral energy coming into it than leaving it, and a negatively charged down-quark (one + anu to two − anu) will have more astral energy leaving it than coming into it. Because of this energy imbalance, up- and down-quarks are attracted to each other in an effort to seek a collective balance in their movement of energy between the astral and etheric dimensions. Since there are three quarks, a combination of up-and down-quarks, in both protons (two up-quarks to one down-quark) and neutrons (one up-quark to two down-quarks), the same energy imbalance occurs within them as well. This concept can also explain the charge in atomic elements and molecules. You can read about my observations regarding forces in quarks and atomic elements in appendix B.

Magnetism

In 2020, I wrote an article that showed how magnetism could result from the movement of energy between the astral and etheric dimensions.[35] One characteristic of the structure of the anu could explain magnetism (Figure 12[36]). If anu serve as a bridge to move energy between two dimensions, then what we call magnetic field lines could result from the cumulative effect of positively-charged anu attracting their negatively-charged counterparts at the opposite end of the magnet. The cumulative attraction between positively- and negatively-charged anu in a magnet could also explain the pull felt when opposite poles of two magnets are placed close together.

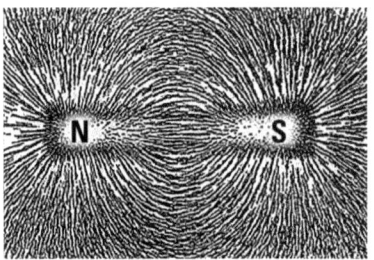

Figure 12: Magnetic field lines
(photo credit in endnote)

Occult Chemistry depicts the iron atom as a jack (Figure 13), similar to the jacks used in a child's ball-and-jacks game. Each arm is packed with anu, shown as dots in the illustration. The trick to making iron

Figure 13: The shape of an Iron (Fe) atom, as drawn in the book *Occult Chemistry.*

or another metal magnetic is to polarize the anu. This means that the anu in the top half of the atom would have one charge and the opposite in the bottom half. When polarized iron atoms join, based on the shape shown in Figure 13, they would form a lattice. The outside ends of this lattice, however, would not have oppositely polarized anu with which to connect. As a result, the anu would have to connect to oppositely polarized anu at the remote end of the magnet, which would create magnetic field lines.

Dark Matter and Dark Energy

Science says that visible matter in the universe makes up less than 5% of the observable universe. This unseen remainder is composed of about 68% energy, also called "dark energy," and about 27% matter, also called "dark matter." Using wave-based physics and the concept of interpenetration, dark energy and dark matter could be explained as matter and energy found in multiple dimensions around planets and other celestial objects that co-exist with us in the physical universe, but at different frequencies, or rates of vibration.

Further Reading – How Different Dimensions Can Have the Same Vibrational Frequencies within Them

In chapter three, we talked about interpenetration and how multiple realities could exist within the same dimension (i.e., the physical dimension) by vibrating at different frequencies, just as different radio and TV station transmissions co-exist unseen all around us. The formula in the text box on the next page shows that a change in a vibration's wavelength (the length of a single wave from one end to the other), will change the frequency at which it is vibrating.[37]

A different section of the same formula shows how multiple dimensions can co-exist in the same space. That component is the ratio T/μ, which is tucked beneath the square root sign on the formula's right-hand side. This ratio shows a direct relationship between tension (T) and density (μ) and how it can affect frequency.

Standing Wave Vibration on a String Formula

The formula, which describes standing wave vibration in a string, is:

$$f = v / \lambda = m / 2L * v = m / 2L * \sqrt{T / \mu}$$

Where,

- f = Vibrational Frequency (vibration of the string)
- v = Velocity (speed at which the wave travels)
- λ = Wavelength (length of a single wave)
- m = Harmonic Number (a positive integer where "1" represents the lowest standing wave vibration frequency)
- L = Length (length of the vibrating string)
- T = Tension (tightness of the vibrating string), and
- μ = Linear Mass Density or Mass per Unit Length

   (diameter, length and material composition of the string)

According to the formula, if we change the tension (T) and the density (μ) to move to a new dimension but keep the ratio of tension to the dimension's density the same—such as a ratio of 2:2, 5:5, or even 19,762:19,762, then the frequency range WITHIN that dimension will remain the same as the one we just left. This implies that multiple dimensions can co-exist in the same physical space and they can each maintain the same frequency range within the dimension by ensuring that the tension/density ratios remain in the same proportion.

To visualize the ratio between tension and density, think of the different dimensions as separate guitar strings (Figure 14). If you have ever played a guitar, or closely examined one, you will see that each of its six strings varies slightly in thickness (they each have a different "density"). Each string will play a different note when plucked; but when we shorten the string's length (which changes the "tension") by pressing it down on a fret on the neck of the guitar shown in Figure 14, and likewise press the next string on certain fret, we can play the same musical note (the same frequency) on a different string, which is similar in concept to how multiple dimensions can co-exist.

To the right of the guitar neck in Figure 14 are the names of each dimension discussed in chapter three. The string at the bottom, the thickest of them all, represents the physical dimension, our physical universe, the densest of all the dimensions. As we move up through the "dimensions" (the strings), their diameter gets smaller (less dense), so to produce the same "note" as we step up through increasingly less-dense dimensions, we must proportionally decrease the tension by using a different fret (i.e., the "E") on the guitar neck to effectively make the string longer.

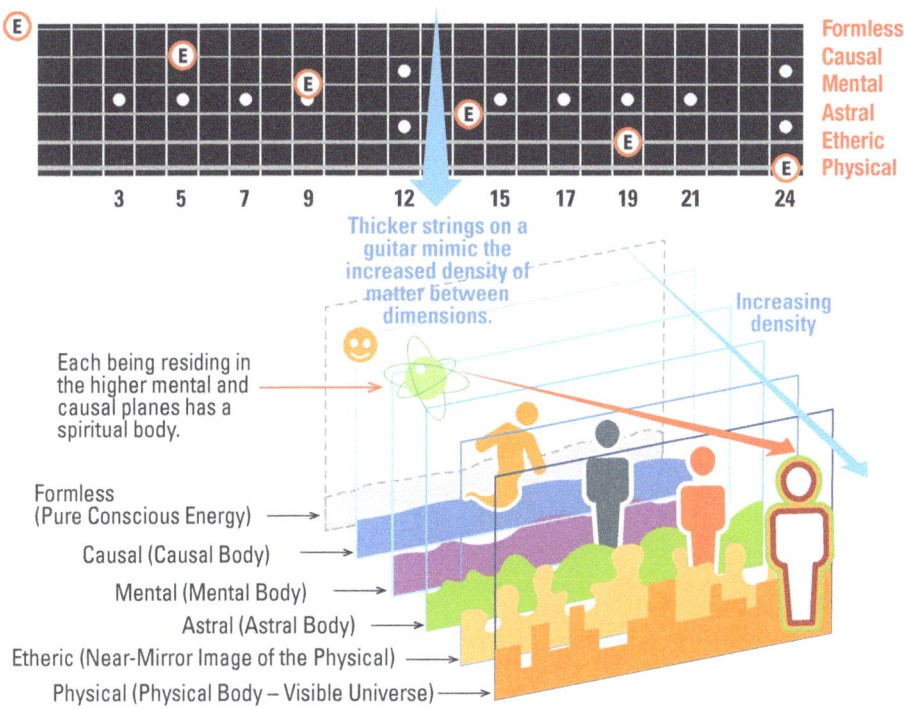

Figure 14: Dimensions have varying densities

CHAPTER FIVE
Explaining Miracles in the Bible with Wave-Based Physics

> **"It is enough for the present to say that parapsychology has in a real sense confirmed the spiritual (i.e., extra-physical) nature of man."**[38]
>
> —Dr. J. B. Rhine, founder of the Rhine Research Center

In this chapter, I explain how wave-based physics can explain miracles recorded in the Bible. The section headings in this chapter are the same as those in appendix E, which contains a comprehensive list of miracles in the Bible. The key point I want to make here is that miracles in biblical scripture are paranormal phenomena that are no different than those observed today! This goes back to my comment in the Introduction that all actions in the universe—past, present and future, obey the same laws of physics. They merely are called by different names!

I contend that miracles recorded in the Bible are really spiritual abilities that beings in non-physical dimensions use to communicate, travel and manipulate objects. To add support for this idea, Dr. J.B. Rhine, an early parapsychologist at Duke University and the person who coined the term extrasensory perception (ESP), proposed in the late 1970s a new area of research he called the Parapsychology of Religion.[39] In his paper, he concluded that miracles described in the Bible were known parapsychological phenomena. In 2021, Brazilian parapsychologist and professor Everton de Oliveira Maraldi published the book *Parapsychology and Religion*, in which he discusses similar topics.[40]

The miracles recorded in appendix E are grouped under five headings: telepathy, remote perception, energy healing, psychokinesis and non-local consciousness.

Telepathy (Communicating Images or Thoughts Between Two or More Beings) When a non-physical being "speaks" to someone on earth, it is important to consider that this communication is non-verbal.

Regarding controlled studies of telepathy, in 2016 there had been 4,674 sessions in 122 experiments performed at twenty labs since 1920. A participant success (hit) rate of 32 percent has been stable against chance (25 percent) for the past twenty years. With the breadth and time span of this research, the probability that the observed effect is a chance result is one in 300 trillion quadrillion (3×10^{-30}).[41]

You must have a quiet mind to be able to sense and discern these thoughts as not being your own. An example of how telepathy can be used in daily life is described in the book *Mutant Message Down Under*, in which the author Marlo Morgan lived with an aboriginal tribe in Australia.[42] An interesting comment she makes at the end of one chapter is that telepathy didn't work for her as long as she had something in her head or heart that she felt necessary to hide. Access to telepathy required her to be at peace with everything.

Remote Perception (Projecting One's Conscious Energy to Perceive Remote Surroundings) This is a short but important list. Remote viewing is an ability that projects one's active consciousness to a remote location to observe what is there. In a wave-based world where consciousness can function independently of a body, remote viewers could either sense remote events or project their consciousness to a remote location to obtain information. This ability has been implied in *Star Wars* movies, where its actors noted that they felt "a disturbance in the force."

There are a number of remote viewing books available on the market. There is even a remote viewer's association—IRVA, The International Remote Viewing Association (www.irva.org), dedicated to promoting the development and responsible use of remote viewing. Edwin May and others similarly describe how the US and Russian governments both used remote viewing to spy on each other in the book *ESP Wars East & West*.

Energy Healing (Healing a Person's Etheric Body, Which in Turn Heals Their Physical Body) There are a number of different ways in which a person's body can be healed, as indicated in chapter twelve's further reading. Note that these methods are vibrational and energetic, and that healing could be done remotely (via quantum entanglement) and through an object (e.g., a handkerchief) as easily as through a person. All of these methods have been demonstrated in parapsychological studies.

Psychokinesis (The Manipulation or Manifestation of Physical Objects) In a wave-based reality, especially in earth's lower mental dimension where the bodies of objects and individuals have been described as wispy, psychokinesis would be an ideal method for moving objects, and even creating spaces, in that environment.

There were significant psychokinesis events recorded in the Bible. One of the best modern day equivalents was described in Jeffrey Mishlove's *The PK Man*, which tells the story of an individual with powerful psychokinetic abilities who could cause earthquakes and weather events, among other things. In the early chapters of another book, *The Energy Cure*, Bill Bengston, PhD describes how an energy healer he met during his early twenties was able to make clouds disappear at will.

Jesus's feeding of more than five thousand people with only five loaves of bread and two small fish was a paranormal event. If we assume that food at its most elemental level is nothing more than a complex waveform, then one can imagine how a "loaves and fishes" waveform could have been replicated by Jesus—similar to how the starship Enterprise's replicator in *Star Trek* episodes did it.

Non-local Consciousness (Accessing Spiritual Abilities or Traveling Outside the Physical Body) In a particle-based world, the ability to travel outside of one's body implies the need for a physical body for the concept to make sense—but not in a wave-based world. This book shows in chapter seven that the "death" of one's body occurs in multiple dimensions, and that bodies should be treated more as vehicles for existence in that dimension instead of as consciousness itself. This makes the concepts of near-death experiences and "bringing the dead back to life" easier to comprehend. If consciousness is eternal and can exist in multiple bodies, then to return it to a body that it had once inhabited can be more easily understood.

Further Reading – A Wave-Based Explanation of the Double-Slit Experiment

In a famous scientific test of the composition of matter, the Double-Slit Experiment, photons[43] of light are fired at a metal plate containing two extremely narrow slits. After the photons pass through the slits, they strike a photographic plate that reveals their impact points. This experiment has been replicated hundreds of times with the same results, all of which show that when photons are not observed, they produce a *wave-like* ripple pattern on the photographic plate. However, if the photons are observed—to determine which slit they pass through, for instance, the photons will form two *particle-like* stripes when they strike the plate. The theory of Wave-Particle Duality, which says that matter in its smallest forms can exhibit properties of both waves and particles, was created to explain this experiment's unusual results.

Now consider this *wave-based* thought experiment. Assume that multiple realities exist within earth's physical dimension, co-existing in the same space but each vibrating at different frequencies - just as multiple FM radio stations co-exist in the same physical space but broadcast on different frequencies. Now consider light coming from the sun. Assume that the sun emits this light in the etheric dimension, not the physical dimension. We can see the sun's light in the physical dimension because the etheric dimension is so closely tied with the physical. Now assume that the sun's light, traveling through the etheric dimension, has NO specific vibration. This enables it to be seen by EVERY reality on earth in the physical dimension and explains how light can have wave-like properties. When light is observed in our reality, the act of observing it in the physical dimension imposes our reality's SPECIFIC VIBRATION onto the light, similar to how a carrier wave creates a steady vibration at a specific frequency, which enables an FM radio station's broadcast to travel through the air at that frequency. The imposition of a specific vibration onto light when it is observed explains how light can have particle-like properties.

CHAPTER SIX
Explaining Paranormal Abilities with Wave-Based Physics

Most people view the paranormal as taboo dark arts we should have nothing to do with. This makes no sense because these phenomena are a natural part of our world, including the non-physical world in which we will one day find ourselves. I believe that in the future we will actually use these abilities as part of our normal day-to-day existence. One reason people are reluctant to embrace, or even consider, these abilities is that they don't understand how they work. The answer is that it's wave-based physics! Below are examples of some of these "other-worldly" phenomena hidden just beyond our common perception.

UFO Appearance, Speed and Maneuverability

The appearance and disappearance of UFOs (Unidentified Flying Objects) is purely a function of wave-based physics. If you vibrate a craft at a frequency beyond the frequency in which we live, the craft will disappear. To make it reappear, change the vibration back to our frequency.

I hinted at what I think is the source of a UFO's speed and maneuverability in the further reading at the end of chapter two. If you can stop the spin of an atom's quarks, you can stop the angular momentum creating the sense of mass in that object. This may be how UFOs maneuver, moving through the air at great speed. If a UFO needed to make a sharp turn, the pilot would slowly activate quark spin in a particular (i.e., X, Y, or Z) direction, which would create the angular momentum that would rapidly bring the craft to a halt, enabling it to quickly change course.

Stopping the spin of quarks in all directions may also render a craft impervious to attack with mechanical weapons. A missile, for example, would pass right through the craft because there was no mass creating resistance to stop it—the craft would only be a waveform.

Unseen Assistance— Invisible Helpers

This work is an application of telepathy, remote perception and going out-of-body at will, and how it can be of benefit to humans living on earth. Charles Leadbeater wrote a book titled *Invisible Helpers* in 1899 (revised in 1915) in which he described events where people claimed to have been helped by angels, gurus living thousands of miles away, or mothers who had passed years before.[45] The book goes on to talk about how some living humans do similar things today. From a technical standpoint, an invisible helper needs the ability to go out-of-body at-will as well as the ability to either project an image to other humans or temporarily materialize their wave-based etheric body while their consciousness is in a remote location.

One practical application of invisible helpers is *soul retrieval*.[46] People who participate in this act go out-of-body to escort humans

Why are Some Aliens So Short?

The two coins in Figure 15 were once the same size. The coin on top shrank after it was subjected to a high-strength electromagnetic field.[44] Some scientists believe that UFOs may be powered by magnetic fields. If this is true, the strength of those fields may explain why we see images of short "grey" aliens in UFO literature—or it could alternatively lend support to a new UFO conspiracy theory!

Figure 15: An electro-magnetically shrunken US silver dollar

who have died, but are either afraid of leaving earth or don't believe they have died, to non-physical beings in the afterlife who help them readjust. A variation of this work is called *soul release,* or the act of helping people die, which is just as beneficial. One example of this work can be found in the book *Drop the BS (Belief Systems) and Be,* in which author Keli Adams writes about her soul release work. She was "tasked" to go out-of-body and appear to people aboard the Dashun, a Chinese ferry that sank in the Yellow Sea; sailors aboard the Russian submarine Kursk after it sank in the Barents Sea; and people jumping from the North Tower of the World Trade Center in New York City on 9/11.[47] Her etheric body appeared to these people as whatever they needed to see—Buddha, Jesus, Muhammad, an angel—to help them accept their impending death without fear and know it would be alright.

Mediumship and Channeling

Mediumship is inter-dimensional telepathy. Through the use of an intermediary, humans in the physical dimension can communicate with non-physical beings in the astral. Telepathy is how non-physical beings communicate with each other.

Channeling is another form of inter-dimensional communication that's undertaken when non-physical beings have moved into dimensions beyond the astral, especially when the being has dropped its form-based bodies. Channeling is a technique where an intermediary (a medium) allows a non-physical being in a higher dimension to use their own body to communicate with people in the physical. While this is happening, the medium is in a trance state and usually cannot remember what transpired or what was said.

Further Reading – Prophets, Mediums and the Ethical Use of Spiritual Abilities

Why does mediumship have such a negative reputation in the Bible? The primary passages that relate to mediumship in the Hebrew Bible (Old Testament) include: Leviticus 19:31,

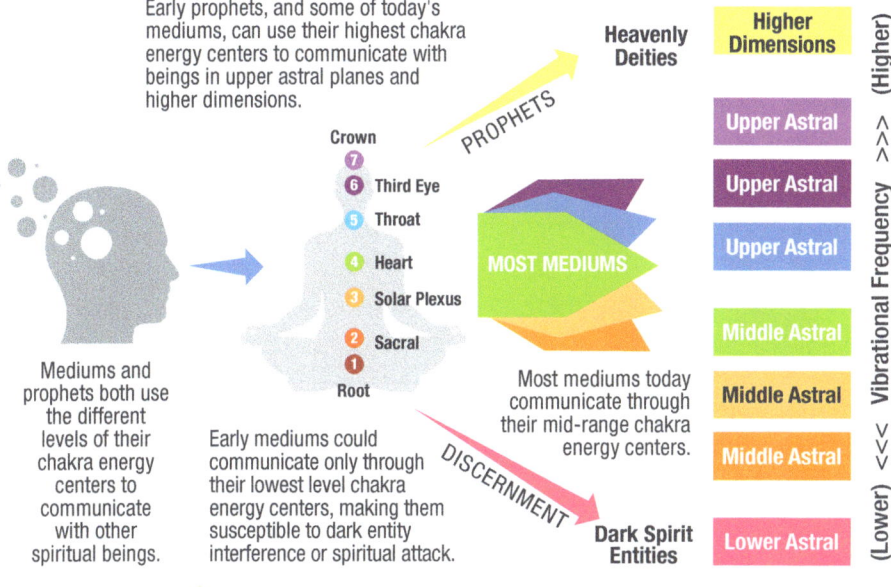

Prophets versus Mediums: The Use of the Chakra Energy Centers for Spiritual Communication

Early prophets, and some of today's mediums, can use their highest chakra energy centers to communicate with beings in upper astral planes and higher dimensions.

Crown
7
6 Third Eye
5 Throat
4 Heart
3 Solar Plexus
2 Sacral
1
Root

PROPHETS

MOST MEDIUMS

DISCERNMENT

Heavenly Deities

Most mediums today communicate through their mid-range chakra energy centers.

Dark Spirit Entities

Mediums and prophets both use the different levels of their chakra energy centers to communicate with other spiritual beings.

Early mediums could communicate only through their lowest level chakra energy centers, making them susceptible to dark entity interference or spiritual attack.

Higher Dimensions

Upper Astral

Upper Astral

Upper Astral

Middle Astral

Middle Astral

Middle Astral

Lower Astral

(Higher) >>> Vibrational Frequency <<< (Lower)

Figure 16: The difference between prophets and mediums

Leviticus 20:6, Leviticus 20:27 and Deuteronomy 18:9–13, and Deuteronomy 18:14–15. These passages mostly say the same thing: do not listen to ghosts or familiar spirits, but Deuteronomy 18:14–15 goes a step farther by telling people that prophets will be provided from among the Israelites and that the people should listen to them.

So what advantage did prophets have over mediums when the Hebrew Bible was written, between two thousand and three thousand years ago? Prophets and mediums are both telepaths. Figure 16 provides an explanation. Three thousand years ago, mediums did not have the high levels of consciousness that many mediums today have acquired. As a result, when they used their telepathic skills, they could only communicate with non-physical beings in the lower astral levels, including the lowest astral level where dark entities could trick people and lead them astray. Prophets, on the other hand, had much higher developed consciousnesses and could communicate telepathically with beings in the highest levels of the astral dimension, and channel beings from even higher dimensions.

At that time, the best information from the non-physical world came from prophets. Today, when people have much higher levels of consciousness overall, a number of mediums can speak telepathically into the upper astral levels and channel beings who exist in even higher dimensions. In the Greek Bible (New Testament), 1 John 4:1–6 provides good counsel—to not believe every spirit, but to test them to determine where (i.e., from what astral or higher level) they come from.

Generally, people who use their spiritual abilities for the benefit of others (such as healing) or for the "greatest and highest benefit for everyone concerned" are people using these abilities for a good purpose—to help humanity. If you hear someone talk about these abilities as a "power" to be acquired, or if they want to use them to benefit themselves, such as winning the lottery, to get revenge or make someone fall in love with them, this is generally considered a dark use of these abilities.

Another way to discern how people are using these gifts is to observe their behavior. Do they act with humility, treating others as being better than themselves? Are they interested in acquiring power, influence or celebrity, or do they try to deflect attention away from themselves? Do not believe a spirit—or a person claiming to have access to a spirit—without first testing to determine its disposition.

CHAPTER SEVEN
Explaining Death with Wave-Based Physics

Based on what we have learned about wave-based realities and interpenetration, it becomes easier to imagine how our consciousness, in a non-physical spirit form, can pass from the physical dimension into the etheric dimension (which interpenetrates it) and ultimately pass to the astral dimension, where it will remain between its excursions to earth. An example of this movement is the concept of death. When we "die," our physical and etheric bodies—the ones with which we are most intimate and familiar—are shed and left behind because they are too dense to enter the less-dense astral dimension.

- **Appendix A** provides observations that support the claims that physics is wave-based, that multiple wave-based dimensions exist, and that these dimensions are inhabited. The appendix also suggests experiments to disprove those claims.

- **Chapter Three** introduces earth's wave-based dimensions and includes facts about each.

- **Chapter Three** also shows (via a thought experiment) how a wave-based consciousness can move down through multiple higher dimensions to inhabit a human body in the physical dimension. In essence, it demonstrates how our eternal consciousness can reside in another dimension and descend to live a human life in the physical dimension.

- **Chapter Four's Further Reading** explains how the different dimensions have different densities and how more dense bodies are shed when a consciousness moves into a less dense dimension. This point, along with the items above, explains what humans perceive as bodily death.

In summary, since the astral dimension is less dense than the physical dimension, WE MUST LEAVE OUR DENSER PHYSICAL BODY BEHIND when our consciousness moves into

this higher dimension! This is what happens when we die; we "shed" our physical body.

Theosophical literature clearly describes the shedding of bodies as consciousness moves between multiple wave-based dimensions. I include the book and page numbers to make these references easier to find:

- Physical body death – The withdrawing of the etheric body, along with prana (energy) – Powell, *The Etheric Double*, chapter XVI, page 70
- Etheric body death – Powell, *The Etheric Double*, chapter XVI, bottom of page 71
- Astral body death – Powell, *The Astral Body*, chapter XXIII, pages 204–206
- Mental body death – Powell, *The Mental Body*, chapter XXX, pages 261–262

Additionally, Arthur Powell's *The Causal Body and the Ego* provides an excellent overview of the process. Charles Leadbeater's *The Devachanic Plane*, pp 78–79 also mentions this process.

CHAPTER EIGHT
Does Infinity Exist in Nature?

Physicists use mathematics to predict the existence of previously unknown properties in nature, but what happens when physics rejects the math? Consider this paradox: The concept of infinity is regularly used in mathematics, but it violates physics laws of the Conservation of Mass and Energy, which state that in a closed universe, mass and energy cannot be created or destroyed. If infinity goes on forever, then an infinite quantity of whatever is being measured can be created, and because of this *contradiction*, science does not recognize infinity as existing in nature. I want to show in this chapter how infinity COULD exist in nature, but it calls into question the assumption that reality consists of only three spacial dimensions.

Infinity's Existence in Four Dimensions

Consider a tesseract, the object in the upper portion of Figure 17. A tesseract is a fourth-dimensional cube. It resembles a cube within a cube when viewed in three dimensions. The unique thing about a tesseract is that the cubes within it can move![48] The fourth dimension is the first dimension in which a shape can loop back upon itself. In a tesseract, the inner cube's corners rotate out one side, then loop back upon themselves by expanding outward and contracting inward, before returning to its original position. This movement can be

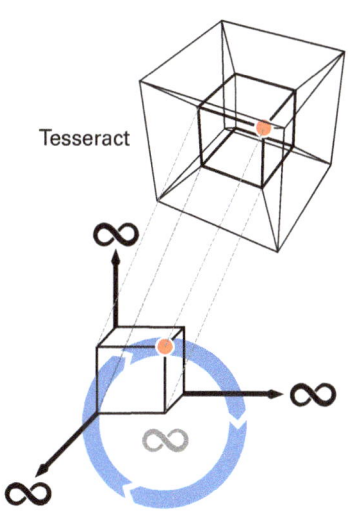

Figure 17: The tesseract and infinity

hard to visualize, which makes the tesseract movement animation in the endnote above helpful to review.

In searching for a mathematical representation for this fourth-dimensional looping property, infinity immediately came to my mind, because looping can go on forever. In three-dimensional space, infinity does not loop; it exists as either a string of increasingly larger whole numbers or as a string of increasingly smaller fractions. In the fourth dimension, though, I propose that these two infinity "strings" join together, forming a loop that rotates back upon itself, just as a tesseract loops! This explanation shows how infinity could exist in nature without violating the laws of physics, but is there an actual example of its existence in the universe? Yes. I propose that there is - the property of scale.

Scale as an Example of Infinity in Nature

Scale overarches length, depth and height and provides perspective from a specific point in space. Scale is not a point. It's more like an elevator you can ride inward or outward from any point. For example, pick a point in space, then imagine hopping on the scale elevator for a ride. Its first stop is the bottom floor, where you get an overview of the very smallest subatomic particles that exist at that point in space. Then take the scale elevator to its opposite extreme and this time observe the magnificence of the entire universe from that point's unique perspective. If the scale elevator stops between these two extremes, you can see the objects that exist at that level of scale. An example of the scale elevator and its many stops along the way is the web-based application *Scale of the Universe2*.[49]

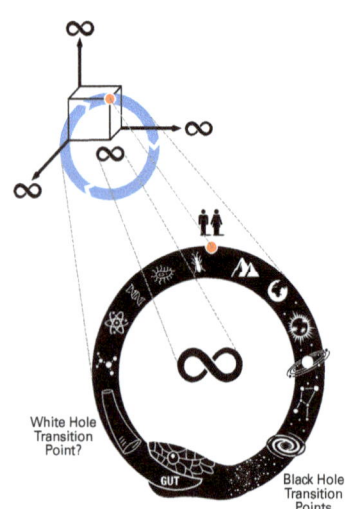

Figure 18: The Cosmic Ouroboros as representative of a fourth-dimensional infinity loop

In the 1980s, Sheldon Glashow, a 1979 Nobel laureate in Physics, presented the idea that micro- and macro-scale objects could loop, but he did not extrapolate this concept to fourth-dimensional space. To illustrate his point, he chose the ancient symbol of a serpent eating its tail to describe the looping of both micro- and macro-scale objects in nature, dubbing it the Cosmic Ouroboros[50] (See Figure 18). The tip of the tail represents the point at which the smallest and largest scale objects joined, a point where Glasahow said gravity was the dominant and controlling force.

In the early 2000s, astrophysicist Joel Primack and philosopher Nancy Ellen Abrams extended Glashow's Cosmic Ouroboros drawing by predicting the existence of dark matter[51] in a void at the micro-scale where no matter was known to exist.[52] In this book's discussion of scale, Primack and Abrams' dark matter is replaced with a black and a white hole, as illustrated in the area on either side of the mouth of the Cosmic Ouroboros in Figure 18. (You will have to imagine this …)

Black holes exist at the galaxy and supercluster levels and also may exist in the observable universe.[53] White holes, the theorized opposite of black holes, could be points where matter from black holes is recycled back into the universe. White holes might also be the source of galactic cosmic rays, which are composed of particles that include atomic nuclei and electrons. If white holes do recycle matter from black holes, then this would complete scale's fourth-dimensional infinity loop (as well as span the Cosmic Ouroboros's mouth in Figure 18 to complete the serpent's infinity loop)!

Now consider this: if it can be shown that we live in four spacial dimensions, that infinity exists in nature, and that the dimension of scale loops at the point where black and white holes join together...could what we call the Big Bang simply be the emergence of matter from a white hole? Let me say it again. Could the Big Bang simply be the emergence of matter from a white hole? If so, we can now say that we understand its place within nature as well.

We just showed how different scales can be linked together figuratively, but can they also be linked together physically? I propose that they are, using electromagnetic forces in the shape of a torus, a circle of energy with the added dimension of depth: a donut shape. Most pictures of a torus are either hand-drawn or digital illustrations. Typically, the torus's internal structure is not shown, nor do these illustrations convey enough information to discern how a torus can stay together. However, 3D-printed models of a torus do exist (See Figure 19), and

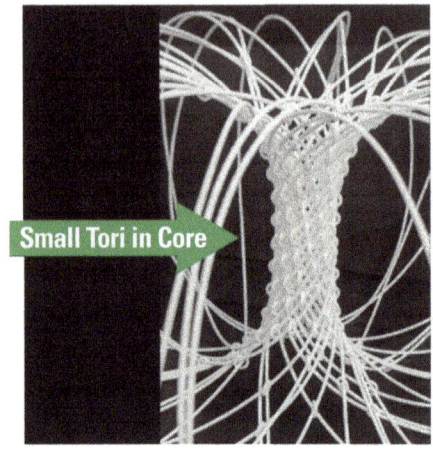

Figure 19: Small tori in the core of a 3D-printed torus

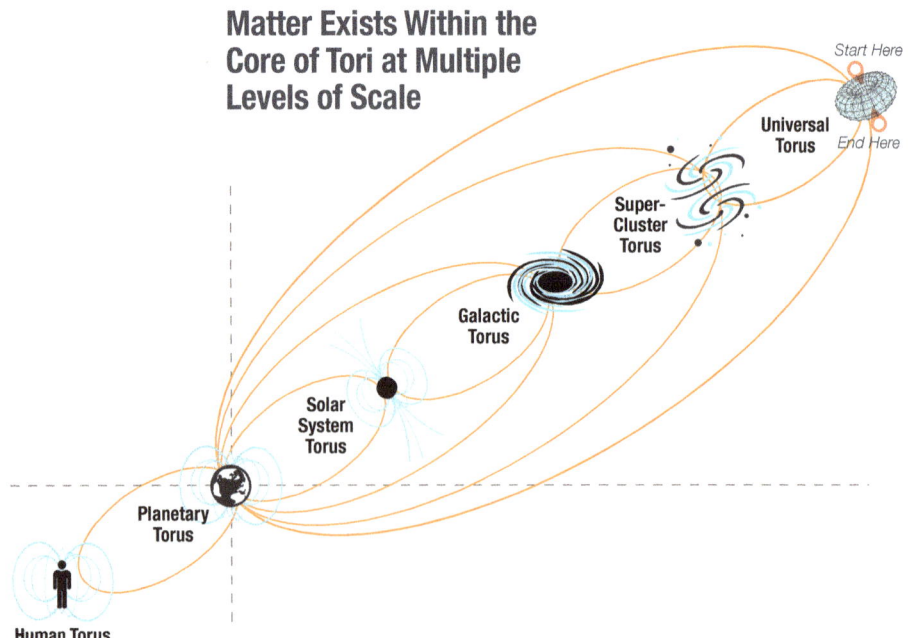

Figure 20: Matter exists within the core of tori at multiple levels of scale

they do show how the torus could maintain its structural integrity. An examination of this structure is enlightening.

The photo in Figure 19 is focused on the inner core of a 3D-printed torus. The core of this torus consists of a number of small circles—smaller tori (tori is the plural of torus)! Seeing this, it becomes easy to imagine how the structural integrity of a large torus could be dependent upon the integration of multiple smaller tori within it. This could mean that tori of various sizes—like the torus that surrounds a person—could be integrated at some point with the torus that surrounds the planet earth, and that these multiple smaller tori contribute to the structural integrity of a larger single torus. If correct, the torus shape could be the connector for scale from the very smallest particles to the very largest structures in the physical universe, as shown in Figure 20. This characteristic would not only show how entities in the universe could be linked but also demonstrate how human existence on earth could be tied to the existence of the universe itself. I discuss this in greater detail in appendix C.

Further Reading – Death as a Fourth-Dimensional Process

Flatland: The Movie[54] is an animated story in which a three-dimensional object passes through a two-dimensional world (Figure 21). As demonstrated in the story, a three-dimensional sphere passing through a two-dimensional world would look (from above) like a two-dimensional CIRCLE that would appear, grow, shrink, then disappear.

Figure 21: A 3D sphere passing through a two-dimensional world would look like a growing and shrinking CIRCLE

Figure 22: A 4D sphere passing through a three-dimensional world would look like a growing and shrinking **SPHERE**

Similarly, if a fourth-dimensional sphere-like shape were to pass through our three-dimensional world, it would look like a three-dimensional SPHERE (Figure 22) that would appear, grow, shrink then disappear.

This pattern parallels what we experience on earth as the pattern of birth, life and death (Figure 23). Could this pattern suggest that human bodies have a fourth-dimensional component to them? Could this component be the astral dimensional form that our human body takes while it is living on earth?

Figure 23: Progression of birth, growth, maturity and Death

STRUGGLE II — EXPLORE IT

The best way to explore somewhere new is to research that destination beforehand and then create a plan that ensures you visit sites of interest, ones that may have special meaning for you. The same is true when exploring the spiritual world. Otherwise, the saying that "If you don't know where you're going, you'll probably end up somewhere else" comes into play.[55]

This section helps you prepare for your initial spiritual experience(s) and suggests developing a spiritual development framework to chart (and help recall) your journey. After a personal plea to not explore on your own, which is where even angels fear to tread, I talk about the concept of spiritual evolution and how similar it is to its more familiar sibling, physical evolution. It ends with a discussion of spiritual abilities, healing and how the non-physical world can act on your thoughts and make them real.

CHAPTER NINE
Preparing for Spiritual Experiences

One goal of spiritual development is to establish communication between humans here on earth and their non-physical helpers, beings who seek to teach us how to become more loving and peaceful (and certainly less warlike). The idea that spiritual entities exist (and may want to help us earthlings) is a most challenging concept to digest—in fact, based on my own experience, it can take years to conclude that there is a non-physical world. Once a spiritual connection has been established, however, and your awareness of these non-physical entities grows, your consciousness will begin to expand.

Below are things you can do to facilitate having spiritual experiences. They assume the user is starting from zero and has not decided whether the spirit world exists. Since it can take months or even years to work through these steps, they should be considered only as a guide.

1 Understand a faith tradition – One of the best ways to develop an understanding of the non-physical world is to participate in a faith tradition. Become familiar with its scripture, gaining both a literal and an allegorical understanding of its passages. Participate in its rituals and acts of service, paying special attention to how these things relate to the spiritual side of the faith. This work can be done in parallel with the steps that follow.

Jibril Caliph Al-Sadat, my internet podcast co-host, has a great analogy for this step—learning how to row a boat on a river. When you first begin learning how to row, your "boat" will be tied close to a dock (your faith tradition). Taken together, your faith's rituals and services will start building an initial comprehension of the spiritual world—like learning to handle the oars of that boat. Once you understand its teachings and have acquired a bit of perspective, you can untie your spiritual boat from the dock, cast off into the river and start exploring the spiritual life on your own. Will you be an expert at it? No, but you will have acquired (and be acquiring) basic spiritual knowledge that will help you better understand experiences to come.

2 **Quiet your mind** – The best preparation for a spiritual experience is to learn how to concentrate—to focus your attention. The goal is to stop the incessant chatter inside your head! It is only when you have a quiet mind that you will be able to differentiate spirit world messages from the running commentary produced by your brain.

   I can't emphasize enough the importance of this work. A quiet mind provides the setting—the velvet background—for your spiritual experiences. It takes time to achieve, so don't become impatient. There are a number of books that teach how to quiet one's mind. Here are two of my favorites:

   ■ *Concentration: A Guide to Mental Mastery* by Mouni Sadhu

   ■ *Concentration: An Approach to Meditation* by Ernest Wood

3 **Develop your internal energy** – Learn how to feel and then manipulate the energy moving within your body. One way to do this is to perform qi gong exercises. Qi gong is a spiritual practice that uses breathing techniques, sound, self-massage and focus to improve both mental and physical health. These exercises also can improve your ability to manipulate the energy in your body and are best done under the supervision of someone experienced in this area. If you don't believe me, reread the chapter "Fools Go Where Angels Fear to Tread," because it's easy to get into trouble without a guide.

4 **Become aware of the different types of non-physical communications** – When your mind becomes quiet, you will be better able to:

   ■ Discern thoughts that are not your own.

   ■ See scenes or videos that play inside your head.

- Sense non-verbal (i.e., telepathic) communication of words, phrases or images.
- Feel special attention being drawn to a particular object or scene.

5 Learn to not fear the unknown – Fear of the unknown is the greatest obstacle to spiritual development and learning to overcome it comes only with experience. Don't try to force spiritual experiences, and when they do happen, don't try to assign a meaning to them. Trust that nothing bad will happen to you and let it happen—don't become anxious or afraid. One way to mitigate fear is to treat the exploration of the spirit world as an opportunity to play and discover. Approach it the way a child does when they are learning something new.

Seek Direction

Once you find a faith tradition that resonates and begin to quiet your mind, you'll want to plan how to pursue your spiritual journey. A good spiritual development framework will show you how far along you are and point to where you will want to go. Think of it as a guide for the road ahead.

Spiritual Development Frameworks

A number of such frameworks exist, many of which are designed to recognize the uniqueness of each individual's journey. The framework I developed for my own spiritual journey (Figure 24) has at its core a spiritual path described in *The Critical Journey: Stages in the Life of Faith*.[56] A good supporting work for this framework can be found in the book *Move: What 1,000 Churches Reveal About Spiritual Growth*.[57] A mega-church in the Midwest sponsored this survey, in which over 250,000 people from over 1,000 church congregations were asked about their spiritual growth. It lists key success factors church leaders can use to cultivate spirituality within a congregation and points out some pitfalls to avoid.

Older mystical itineraries are listed in Bernard McGinn's book, *The Essential Writings of Christian Mysticism*.[58] It contains six different spiritual development frameworks, each written by an early mystic. These "mystical itineraries" include:

- *The Threefold Way* (from multiple authors) that consider:
  - meditation, prayer and contemplation
  - animal, rational and spiritual life > body, soul and spirit
  - moral, allegorical and anagogical > purgation, illumination and perfection
- *The Four Degrees of Violent Charity*, by Richard of St. Victor
- *The Mind's Journey into God*, by St. Bonaventure
- *The Mirror of Simple Annihilated Souls*, by Marguerite Porete
- *Sermon 39*, by John Tauler
- *The Ladder of Perfection*, by Walter Hilton

Evelyn Underhill's classic book *Mysticism*[59] highlights different mystical states and mentions a number of spiritual development frameworks, including St. Teresa of Avila's *Seven Degrees of Contemplation*. The chapter headings in Part Two of Underhill's book provide an overview of the concepts:

- The Awakening of the Self
- The Purification of the Self
- The Illumination of the Self
- Voices and Visions
- Introversion: Recollection and Quiet
- Contemplation
- Ecstasy and Rapture
- The Dark Night of the Soul
- The Unitive Life

St. Teresa of Avila's framework is similar in some ways to other frameworks but places the Purification and Illumination of the Self in front of Voices and Visions—the literal before the spiritual. Per Underhill, the realization of the spiritual world brings about an introversion, which leads to inner joy. Then, after the dark night of the soul (somewhat similar to *The Wall* in my framework), you will be led to a deeper understanding of your spiritual experiences.

Two Principles to Remember

There are two key principles to remember while you are on your spiritual journey. First, that each person's spirituality is unique. Two people may have similar intellectual understandings based on the same faith tradition, but after passing through *The Wall* (appendix D), it's practically guaranteed that their perspectives will differ because the Wall experience reconciles one's intellectual beliefs with one's unique life experiences. Even so, we can take advantage of these differences by sharing—and listening to—each other's beliefs without judgment, which can accelerate and enrich everyone's personal and collective spiritual journeys.

Second, the spiritual framework outlined in this chapter's further reading provides *the baseline, the initial point* from which to grow your spirituality and will be worthwhile to work your way through. Once you have established some spiritual depth, it will continue maturing as you learn to love other people and persevere in defending what you believe, which builds moral character. This part of your journey can consume your entire life; the positive side effect is that you are strengthening both your spirit and your life. The ultimate result of this journey is spiritual transformation. You may very well find yourself—years later—back at the place where you started but you will have gained a different mindset and worldview! Pursuing spiritual growth creates a more loving person who is better able to embrace life's many challenges.

Further Reading – My Spiritual Development Framework—An Overview

This framework is discussed in greater detail in appendix D, but I wanted to introduce you to it here. The framework is divided into four parts:

1 The Spiritual Path consists of the six steps that form the body of the arrow. If you look in the symbol key at the bottom of the diagram, you will see that the colors mean something as well. You go through the grey-colored steps when your faith primarily consists of what you have studied—what you intellectually

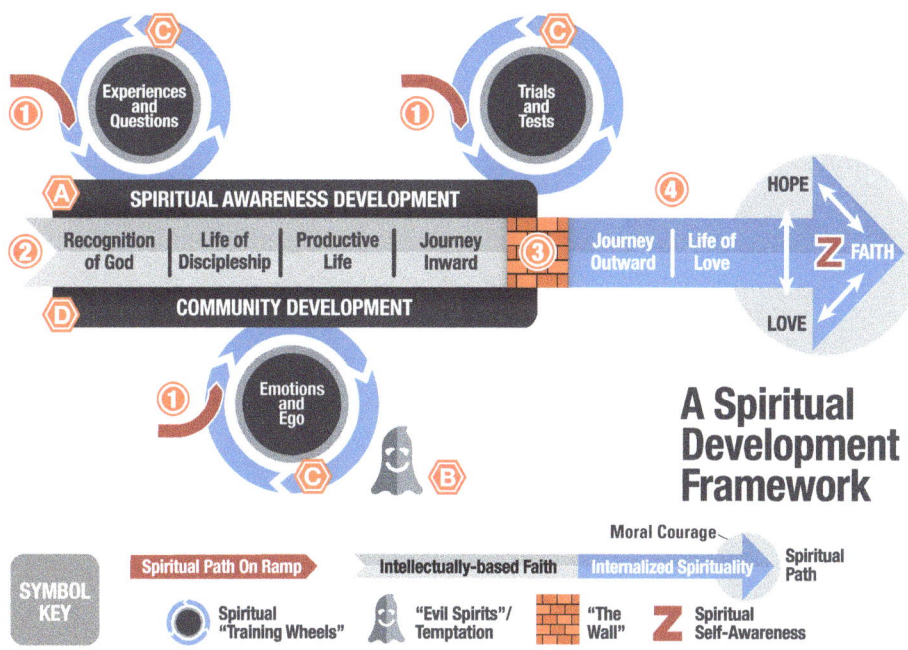

Figure 24: The framework I developed for my spiritual journey

understand—and the blue-colored steps after you have internalized a unique spirituality.

2 Your Spiritual Life surrounds the spiritual path in the main part of the diagram and focuses on developing spiritual practice (Spiritual Awareness Development) and on building spiritual community (Community Development).

3 Life Training Wheels are the three circles that appear above and below the spiritual path. These are things that can throw you off the path. They include:

   a. Experiences and Questions—life events, such as a serious illness or death of a loved one, that leave you asking why this happened.

   b. Trials and Tests—events that affect your everyday life, which could include job loss, a family crisis or problems at your place of employment. It is leaning over the top of *The Wall* because these training opportunities never end—and may become more difficult after your spiritual beliefs stabilize.

c. **Emotions and Ego**—separate from the other two training wheels, this area is focused internally (relates to "us"). Emotional experiences usually come upon us because of temptation (see the smiling "evil spirit" in the chart?) and relate to how we interact with others. In fact, the people around you play a large role in your development here, as they need to forgive you when you make changes to improve your life.

4 **Putting It All Together** (i.e., the tip of the arrow)—love, faith and hope are the products of spiritual maturity. Your life will have improved as a result of walking the spiritual path and using what you have learned will help you better tackle life's experiences, trials and tests.

If you would like to know more, this framework is described in greater detail in section three of my third book, which is listed in the Bibliography.

CHAPTER TEN
Fools Go Where Angels Fear to Tread

The saying "Fools go where angels fear to tread" is a great description for people who choose to explore the non-physical world on their own. This is a mistake—and I say that from hard experience! It's important to find a mentor, a person with a heart for the subject who has traveled the path you are either on or are considering traveling. I started with a mentor but went out on my own within a couple of years, after which some unexpected things began happening to me. I bring this up again only to emphasize the point that having a mentor will significantly increase the probability that you will have an enjoyable experience—and lower the likelihood of serious problems!

The following are examples of things that probably never would have happened to me had I been working with a mentor.

Removing the Unknown

One day, years ago, I concentrated on a rash on my hand in an attempt to command away any spirits that might be causing it. After a few minutes, I felt energy slowly moving down my arms and out through the tips of my fingers. It felt similar to the feeling of slowly pulling an elbow-length glove off your arm. As I kept commanding it, this sensation expanded to cover my entire body. For lack of a better word, I will call what came off my body a sheath. It took me over two hours to completely command this sheath to leave my body: but what happened next was frightening because without its protection, my body became super-sensitive to spiritual activity.

The most disconcerting sensitivity was what I experienced each night in bed: the feeling that 8–10 small spirits were continuously bouncing off my forehead—like ping-pong balls. Each time I tried to shoo these spirits away with my hand, they immediately came back, like moths to a lamp. This happened throughout the night—every night! After three weeks of this torment, I was at my wit's end because I had gotten so little sleep.

A few days later, I was again sitting in my recliner, commanding a spirit out of my body. After this spirit left, however, I felt the sheath I had forced off my body earlier being pulled back over me! I could not sense or see who or what was making this happen. But the "ping-pong balls on the forehead" feeling stopped and I could sleep once again. I felt fortunate to have been the beneficiary of what likely was a true spiritual intervention.

With the benefit of years of reading, I believe what I removed is called an atomic web,[60] a thin shell of energy that prevents the premature opening of communication between the astral and physical dimensions.[61] It can be removed by an emotional shock, alcohol or substance abuse, or even someone trying to force development of their psychic abilities.[62] I believe this web is also removed when someone has a near-death experience, which is why these people often develop psychic abilities afterward. If you do not have a mentor who knows when you are ready to have this "sheath" removed, don't try to remove it yourself or try to force the development of psychic abilities on your own.

Dealing with Spiritual Entities

There are over two thousand different kinds of elementals, a type of non-physical being, and they each do things that are beneficial to our existence.[63] These non-human life forms are alive, sentient and have emotions, thoughts and feelings—just like us. In general, they possess a simpler form of consciousness than humans and come to Earth so that they, too, can learn from the experience and grow spiritually—again, just like us. They are different from the more commonly recognized "nature spirits" such as fairies, pixies and gnomes (yes, they also exist).

Unfortunately, when "my" elementals arrived, I resisted them, primarily out of fear of the unknown. After five years and dozens of attempts to get them to leave, they still kept coming. A useful analogy to describe this situation is the vast number of European settlers that Native Americans encountered during America's period of westward expansion. No matter what they did to resist, the settlers still kept coming.

If you become aware of the presence of spiritual entities in your body, just be curious and experience their presence without attaching any meaning to it. If you choose, you can mentally ask these entities what lesson they are trying to teach you, and you should accept the thought that immediately comes into your mind as their reply. When they do leave, be sure to thank them, with gratitude, for what they taught you. Also, you may want to say that you "release them" when they leave, saying that they are now free to assist someone else.

Embracing Energy Healing

I didn't learn an energy healing technique until years after I had developed a sensitivity to my body's energy. In retrospect, I believe it would have been helpful to have learned Reiki or some other healing modality first. Such an experience would have given me a foundation to help me better understand how energy moved within my body and also would have given me access to another person with whom I could discuss body energy and energy movement.

Developing Energy Sensitivity

I purposely have not discussed the exercises I used to sensitize my body to the energy flowing within it, but I do want to share one lesson from it: be careful how you move energy around your brain. While doing these exercises, I modified one that used energy coming from my palms to move energy around my brain. What I should have done was hold my hands a short distance away from my ears and PULL energy away from one side of my brain with one hand as I PUSHED energy into my brain with the other. Instead, I used both hands to simultaneously push and pull energy into and out of my brain at the same time. After a few weeks of doing this, I began to stutter, and this issue persisted for almost a year. I cannot say whether the way I did these exercises caused this to occur, but it seemed more than just coincidental.

Forcing Spiritual Abilities

Do not attempt to move energy around or near your heart or around your spine. Understanding that you should not mess with your heart should go without saying—your spine, too, because of the energy it carries. Also, there is a ball of energy at the base of your spine, which is called your kundalini. If you mess with either your spine or your kundalini and don't know what you are doing, you risk resetting your body's energy system—or worse. It's best not to touch them.

People are naturally interested in awakening their kundalini, however, there have been many negative kundalini experiences when folks have attempted it on their own, including painful neurological problems that lasted months...or years. The biggest mistake people make is to attempt to directly connect their kundalini to their brain by moving its energy straight up their spine. There is a specific way to make this connection: and it is NOT by moving energy directly up the spine! The best advice - allow your spiritual abilities to develop on their own. **They should not be forced through development practices.**

Further Reading – Spiritual Emergencies—Where to Get Assistance

If you have questions about a spiritual experience or run into problems (e.g., a spiritual emergency) there is an organization that may be able to help you: ACISTE, the American Center for the Integration of Spiritually Transformative Experiences (www.aciste.org). This organization was created by spiritual experiencers for spiritual experiencers. They train mental health professionals, spiritual counselors and life coaches to be sensitive to the needs of spiritual experiencers.

Another good resource is Stan and Christina Grof's book *Spiritual Emergency*. See the Bibliography for additional information.

CHAPTER ELEVEN
Spiritual Evolution

If we assume that we are spiritual beings who come to earth to learn, there should be associated with that learning experience a notable path of growth. In Theosophical literature, this progression in our conscious development involves transitioning from an "animal" nature to a human one, and then to a spiritual nature (Figure 25). Humans are eternal spiritual beings, but because we inhabit temporary human bodies that eventually "die," humanity has come to believe that life on earth is limited and that we are not part of an eternal existence. However, if we embrace the explanation of death that was presented in chapter seven, that our spiritual selves live on after physical death, we can then perceive our existence as part of a process of spiritual evolution.

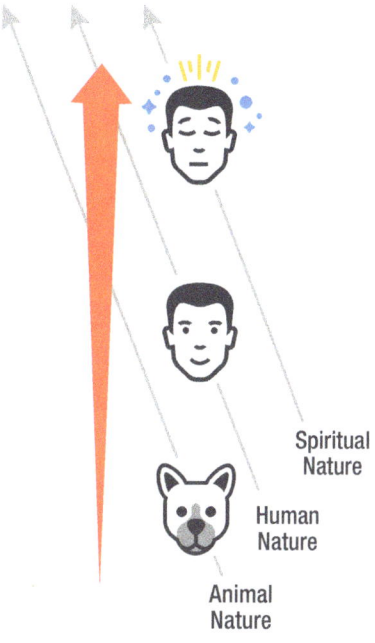

Spiritual Nature

Human Nature

Animal Nature

Figure 25: The process of spiritual evolution

How can spiritual evolution occur on earth? An analogy based on Richard Dawkins's book *The Blind Watchmaker* helps explain the mechanism. In chapter three of his book,[64] Dawkins describes a computer program he devised to simulate the power of physical evolution. His application was designed to see how quickly a computer could randomly produce the 28-character phrase, "Methinks it is like a weasel," a line from Shakespeare's *Hamlet*. The program made changes to a number of randomly-generated text sequences across multiple test iterations. At the end of each iteration, each modified instance was compared to the phrase from *Hamlet* to see if there were any exact matches.

The probability of producing the phrase by random chance is about one in 10^{40} (extremely low). However, when Dawkins introduced an evolutionary function into the program that selected the best match to the *Hamlet* phrase and made it the starting text sequence for all instances in the next iteration, the application was able to duplicate the phrase in about fifty iterations! Dawkins's program shows the power of introducing evolution into an otherwise random chance process to create a dramatic rate of change over a relatively short time span.

This example can also explain spiritual evolution, the process by which spiritual beings mature. Sacred scripture, present in many world religions, can be seen as the objective of earth's spiritual evolutionary process, just as "Methinks it is like a weasel" was in Dawkins's program. Since sacred scripture has essentially remained unchanged for thousands of years, it's enlightening to consider that anyone, in any time period, is able to "plug-in," to connect with that scripture, to embrace its teaching and initiate their own spiritual evolutionary process.

The most important facet of spiritual evolution, though, lies in the act of "inter-generational transfer"—the passing of spiritual knowledge from generation to generation. Going back to the *Blind Watchmaker* analogy, transferring information across generations is the same as taking the instance that comes closest to "Methinks it is like a weasel" in one iteration and making it the starting text for all the instances in the next sequence. Inter-generational transfer enables future generations of humanity to benefit from the experiences of previous generations, which is essential to the rapid rate of growth in the spiritual evolutionary process—and a warning: spiritual evolution could be reduced back to random chance when inter-generational transfer and an objective are not present.

Further Reading – The Power of Your Thoughts

There is a saying in metaphysical circles that "thoughts are things." I'll use Figure 26 and the concept of thoughtforms to explain. It's hard to imagine how thoughts can become a tangible object, but in a way, this can happen. People are generally aware of the theory that humans evolved from animals, specifically primates, but Theosophical literature[65] tracks the growth of consciousness back even farther than that. In fact, it says that what is now our human consciousness started out as much simpler consciousness entities that were embodied in clouds of energy, called "Elemental Essence," in three non-physical dimensions—the higher mental, lower mental and the astral dimensions. It is not until these entities descend into the etheric dimension that they begin to gain some independence—becoming embedded in the material form of minerals, plants, animals and humans. It is my opinion that before conscious entities descend into the etheric, they develop by trying to manifest the thoughts of matter-based life forms.

The chart suggests that plants have no thoughts but that animals and people do. Some researchers, though, have implied that plants do have feelings, measurable electrically when the plant is threatened—or even complimented.[66] However, we can readily imagine that animals have emotions and instincts and that humans can have thoughts that are emotional and egotistical, growing into more reasoned and more abstract thoughts. I'll describe human thought from this point forward.

When we think, our brains create thought vibrations similar to a radio broadcast.[67] When our thoughts interpenetrate the elemental essence in the lower mental and astral dimensions, they are picked up by a conscious entity within that dimension's elemental essence. The conscious entity then leaves the elemental essence as a thoughtform with the intention of making the thought become real. For an analogy of this process, think of the concepts discussed in the documentary The Secret. The chart in Figure 26 basically validates what the film teaches.

The strength of our thoughts determines the strength of the thoughtform. For instance, a vague thought about someone will produce a short-lived thoughtform that will quickly dissipate. However, if we are concerned about someone who has become

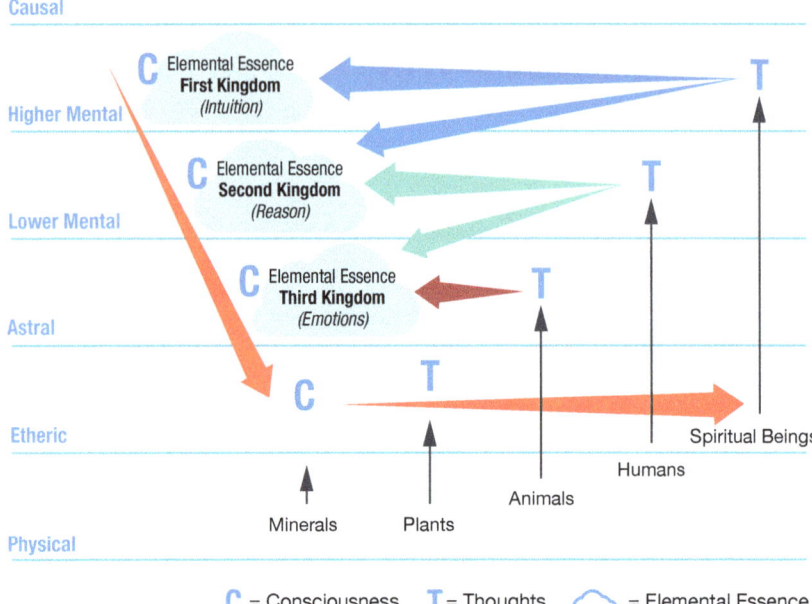

Figure 26: Chart showing how the non-physical world helps thoughts become real

sick, our thoughts will become specific (about the person who is sick), emotional (because of our concern for their welfare) and strong (because of our degree of concern for them). These thoughtforms will last much longer—often hours or days. If we enlist others to pray as a group for this person's healing, our group prayer creates a strong thoughtform whose mission is to help that person become well again.

An example of how our thoughts create reality can be found in the book *The Hidden Messages in Water*. The author, Dr. Masaru Emoto, shows how human thoughts and emotions can affect the way water crystalizes. His book contains a number of color photographs of snowflake-like ice crystals that were exposed to classical music, people saying specific words (e.g., beauty, love, hate, fear) or words that were simply taped to a jar of water.

In one experiment,[68] three glass jars were filled with rice and water. Japanese elementary school children became an active part of the experiment, each day saying positive things, such as "thank you" to the first jar, and negative things like "you fool" to the second jar. Nothing was done to the third jar—either positive or negative. What happened? The jar that received the positive affirmations started fermenting and had a pleasant malt aroma. The jar that received the negative affirmations rotted, but surprisingly, the rice in the jar that was ignored rotted EARLIER than the rice that was being ridiculed! If words can affect water, and the human body is approximately 70% water, imagine what impact our words and actions can have on our bodies and on the bodies of those around us—especially children. Dr. Emoto's book underscores the importance of what we say and do to others.

Up to now, we've concerned ourselves only with the thoughts of an individual, but what about the thoughts of a large group? Think about sporting events, protests, or even the response of a populace to news heard on TV. The collective emotional response of groups creates an especially strong thoughtform called an "egregore." It can exist for a long period of time, especially in locations with regular gatherings, such as European cathedrals over the centuries.

The impact of thoughtforms depends on the receiver's and sender's intentions. If someone sends positive thoughts to someone via a thoughtform, such as for healing, the thoughtform will deliver the healing energy to the recipient. However, if someone sends negative thoughts to someone via a thoughtform, and it has no affinity with the receiver (e.g., the receiver is a nice person), the thoughtform can rebound back and affect the sender. Also, if someone has consistently strong thoughts about something, such as life being miserable, and there is no recipient, that then thoughtform will remain around the sender. This could explain why some people can never seem to emerge from a negative frame of mind.

To summarize, thoughtforms are created when we have strong thoughts about something or someone. We should ensure that our thoughts directed toward others are positive and that we strive to maintain a positive outlook toward our own existence. We should also be aware that our words—as well as our thoughts—have the power to help or harm, especially with children.

CHAPTER TWELVE
Thoughts Regarding Spiritual Abilities

Figure 27 presents a hierarchy of spiritual abilities. (Note that what the diagram labels as "paranormal" I choose to call "spiritual.") No matter the label, these abilities are useful adjuncts to our body's five senses, which become more powerful as we ascend into less dense, higher dimensions.[69] Some people can access spiritual abilities in the physical dimension, although most who can rarely talk about it.

One thing I realized about spiritual abilities while writing this book is that separate groups in the metaphysical world teach the same things—the same techniques—they just call them by different names!

Figure 27: Paranormal abilities are how spiritual beings travel, communicate and manipulate objects.

Many of these abilities require the use of three components: focused concentration, visualization and intention.

People who perform intercessory prayers use them to help those who are sick or injured. People offering the prayer CONCENTRATE on the thought that the person's illness or injury be healed. They VISUALIZE the person being prayed for as being active and healthy, and they set the INTENTION that the prayer is offered to improve the person's welfare.

These same three components are also used to cast spells. The people casting the spell CONCENTRATE on the action or event they want to occur, then they VISUALIZE the action or event happening, and they center their INTENTION on the person who will benefit from it—which could be themselves!

These same three components are mentioned in the documentary *The Secret*[70] and are central to the film's prescription for attaining things you want in your life.

What, then, is happening in these examples, since each of these different techniques must obey the same physical laws? From my perspective, each example creates a thoughtform,[71] which helps our thoughts become things, to bring about what was intended. Our thoughts create our reality. This IS the secret of life on earth, as discussed in chapter eleven's further reading.

There is a purpose for everything we do and experience. I see access to spiritual abilities in the physical dimension as giving humans an opportunity to practice these things in a controlled environment where the risk for damage is limited. We will use these abilities in the future when we achieve a fully non-physical existence—and become busy helping other less developed conscious entities grow and develop.

> **"Human beings have no special sensations in the presence of magnetic fields. Had it not been for the two very contingent facts that there are loadstones, and that the one element (iron) which is strongly susceptible to magnetic influence is fairly common**

on earth, the existence of magnetism might have remained unsuspected to this day. Even so, it was regarded as a kind of mysterious anomaly until its connection with electricity was discovered and we gained the power to produce strong magnetic fields at will. Yet, all this while, magnetic fields had existed, and had been producing effects, whenever and wherever electric currents were passing. Is it not possible that natural mediums might be comparable to loadstones; that paranormal influences are as pervasive as magnetism; and that we fail to recognize this only because our knowledge and control of them are at about the same level as were men's knowledge and control of magnetism when Gilbert wrote his treatise on the magnet?"

—Sir Alister Hardy, marine biologist and founder of the Religious Experience Research Centre[72]

Further Reading – Healing Overview

Before we get started, here are a few things that occur during wave-based healing:

- Healers focus their work on a person's etheric body, not their physical body. Changes made to the person's etheric body naturally propagate to the physical body afterward.
- Healers use quantum entanglement to connect to a person's body when doing remote healing. After the healing session, the healer breaks the entanglement connection.
- The subconscious mind directs vibrational healing in a person's body while they are asleep. A person's conscious mind can use this same faculty to heal themselves while they are awake.
- Thoughtforms are often used, such as during individual or group intercessory prayer, to effect healing of a person's etheric body, with that healing propagating to the physical.

I have observed five categories (the blue circles in Figure 28) of physical, mental or spiritual healing that have been practiced worldwide for millennia. Here's how I believe they fit together. The categories are:

1 Medical Healing – The traditional use of medicine, surgery and other scientifically-proven techniques and therapies to heal a person's physical body.

Five Categories of Physical, Mental and Spiritual Healing

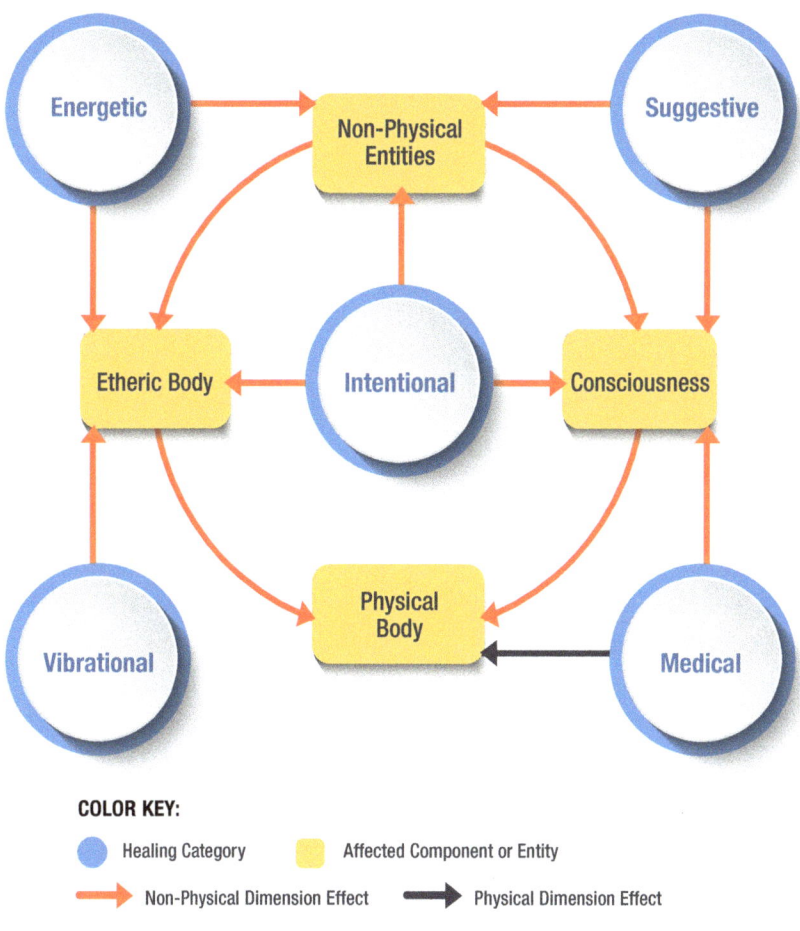

COLOR KEY:

● Healing Category ▮ Affected Component or Entity

→ Non-Physical Dimension Effect ⟶ Physical Dimension Effect

Figure 28: The five categories of physical, mental and spiritual healing

2 Vibrational Healing – The use of light, sound or vibrations to resynchronize wave patterns within a person's etheric body.

3 Suggestive Healing – The convincing of a person's subconscious mind that the physical body has been healed, which causes their subconscious mind to heal the person's body.

4 Energetic Healing – The sending of energy to a person's etheric body to give their body additional strength to heal itself.

5 Intentional Healing – The direct creation of healing thoughtforms, by a group or an individual, and/or the solicitation of higher dimensional beings to heal a person's etheric body. Healers can also invite higher dimensional beings to use the healer's body as an inter-dimensional conduit through which to provide healing to a person.

What Gets Healed

In this section, we will talk about how the healing categories affect a person's body. The Affected Components (the yellow rectangles in Figure 28) include:

1 The Physical Body – This is the domain of traditional medicine, which uses surgery, drugs and physical manipulation of muscle and bones to achieve beneficial therapeutic results. The subconscious mind, which controls involuntary movements in the body, heals the etheric body on a nightly basis, which in turn heals the physical body.

2 The Etheric Body – Healers use energetic and vibrational healing techniques to heal a person's etheric body. Non-physical entities and higher dimensional beings can also effect change in etheric bodies, either through direct intervention or through the use of thoughtforms.

3 Consciousness – Suggestive techniques are used to convince a person's subconscious, which is a separate conscious entity,[73] that the person's body has been healed. The subconscious then proceeds to heal the body.

In medicine, this could be called a placebo effect, where a person's subconscious believes that a sugar tablet, or other pill/treatment, is effective—and it works. Religious healing ceremonies can have a similar effect.

4 **Non-Physical Entities** – Spiritual beings can either cause or cure disease, primarily through working on a person's etheric body. They can also convince a person's subconscious that they created a problem, when they haven't, and if the subconscious believes it, guess what? The person gets sick! Higher dimensional beings or their thoughtforms can also affect the work or the effectiveness of non-physical entities.

Quantum Entanglement and Remote Healing

The way healers perform remote healing is to first connect with the person's etheric body via quantum entanglement. This enables the healer to scan the person's body, and if necessary, perform energetic or vibrational healing from a distance. At the end of the healing session, the healer breaks the quantum entanglement connection with the etheric body of the person being healed.

Wave-Based Healing Diagnostics

There are a number of ways that wave-based healing diagnostic techniques can be used to identify diseases or detect problems in a person's body. These include:

1 **Direct Scanning** – Healers can run their hands over a person to be healed to detect areas of heat, vibration or sparking. The presence of heat implies inflammation. Vibration or sparking sensations imply the presence of non-physical entities or energy balls around diseased or damaged tissue.

2 **Remote Scanning** – Similar to direct scanning, but it is done remotely using quantum entanglement to form a connection with the person being scanned. Healers often use stuffed animals to represent a person's body, which helps them visualize what they feel. A voodoo doll is another example.

3 Trance – When a healer goes into a trance, their subconscious mind can come to the surface. The subconscious mind can communicate and access non-physical abilities and information that is often unavailable to the conscious mind.

4 Astral Sight – Some healers have fourth-dimensional sight, also called astral sight, which is a spiritual ability that enables them to see inside three-dimensional objects. Think of how you, as a three-dimensional being, can see things by peering down from above that a being living in only two dimensions cannot.[74] The same principle applies with 4D sight in a 3D world. This ability enables these healers to see inside a person's body to look for areas that may be diseased or damaged.

Healing Experiments–Visual and Hands On

In the 1970s, two noted psychics, Olga Worrall and Ingo Swann, participated in an experiment to see if healing energy could be seen coming from their hands.[75] The test involved the use of a cloud chamber, a simple device that had earlier been used to prove the existence of cosmic radiation passing through the earth's atmosphere. Results showed that both participants were able to create visible movements within the cloud chamber when healing energy passed from one hand to the other.

Bill Bengston, PhD, has conducted a number of controlled laboratory experiments to test his hands-on healing technique. One of the most exciting claims from this work is the world's first remission of mammary cancer in laboratory mice specifically bred to lack a resistance to this type of cancer. Details of his early experiments can be found in his book, *The Energy Cure*.

Books to Read

This further reading on healing could easily be expanded into a book on its own—and many people have done so. Some books related to vibrational and energetic healing are:

- *The Energy Cure*, by Bill Bengston, PhD, documents a number of controlled laboratory experiments testing the efficacy of the author's hands-on healing technique. This book documents his early work. His more recent work is contained on his web site (www.BengstonResearch.com).

- *The Power of Eight,* by Lynne McTaggart (www.LynneMcTaggart.com), recounts the author's work with small group prayer and large group intentions (e.g., thoughtforms) and the ensuing results. Her book also discusses an experiment she led in which thousands of participants from around the world set a conscious intention to bring an end to a twenty-five-year-long war in Sri Lanka.

- *Co-operative Healing,* by Ernst Eeman, discusses how the body can heal itself using what Eeman called a relaxation circuit, based on the body's positive and negative polarities. This book also documents seventy-one tests, which can be replicated, in which medicines were delivered energetically through a biocircuit—not orally! The book *Biocircuits,* by Leslie Patten, is a more recent treatment of the subject and of Eeman's work.

- *The Etheric Double,* by Arthur Powell, documents the working of the etheric body, which interpenetrates the physical bodies of humans, animals and plants. This is a good reference book.

STRUGGLE III — EMBRACE IT

The spirituality you develop is unique, so the way you embrace that spirituality also should be unique. The thoughts in the first chapter share how I plan to embrace my spiritual nature, but it leaves room for you to create your own way to embrace it too. The next chapters recap the explanations proposed in this book and describe a model that integrates the creation and development of our wave-based reality. The section ends with a call to action to determine whether physics is particle- or wave-based.

CHAPTER THIRTEEN
Dream Big!

I like to work on big ideas, like climate change, that require large numbers of people and long timeframes to accomplish, because big ideas inspire! This is why I mention climate change here—to address the need to find alternative energy sources that reduce the global demand for fossil fuels. When people consider today's energy alternatives, the most common thoughts revolve around solar power, electric vehicles and hydrogen fuel cells. These solutions are *evolutionary* in nature because they are based on our current particle-based understanding of physics. What I want to do is find a *revolutionary* replacement for fossil fuels using innovations based on wave-based physics.

In chapter three's further reading, I speculate whether the anu's purported movement of anti-matter from the astral to the matter-based etheric dimension could be a viable source of power for the planet. If this energy source exists, we could potentially tap it to meet our day-to-day energy needs. If wave-based physics is proven, then developing an energy-based application based on this "new" science could bring inexpensive energy to all parts of the planet.

Further Reading – Developing a Single Global Worldview

Reflecting on this book's content, I realized that the application of its ideas could unify the disparate worldviews and belief systems that separate science, religion and metaphysics (Figure 29). The concept that physics is fully wave-based, and by extension our reality, could serve as the catalyst to bring about this unprecedented change. If this concept is experimentally shown to be true, it

Figure 29: A single global worldview can bring together science, religion and metaphysics

could result in the development of a single global worldview, and a dramatically more unified, more confident, and more joyful world—one with a level of love and mutual understanding that is difficult to imagine today.

If an organization was created to facilitate this change, it might have the following objectives:

1 Perform the neutron decay experiment described in appendix A, to determine whether physics is fully wave-based. (The results of this experiment should settle any controversy surrounding the book's claims—one way or the other. Scientists test ideas that don't prove out all the time. I fully understand that these claims could face the same fate.)

2 **Contingent on the success of objective #1:** Promote the development of a single global worldview.

3 **Also contingent on the success of objective #1:** Develop a technology that utilizes wave-based physics, such as the multi-dimensional engine idea mentioned earlier in this chapter.

CHAPTER FOURTEEN
A Wave-Based Physics Model of Reality

The comprehensive chart in this chapter demonstrates and explains how wave-based physics "connects the dots" between our physical reality here on earth and hidden realities that exist all around us in other dimensions. It also reaches back into time, showing how these interrelations developed and organizes the propositions presented in this book into an internally consistent, wave-based explanation that I (am only somewhat hesitant to) call an alternate model of reality. The four component charts cover the:

1 Formation of earth's dimensions.

2 Balancing of energy between dimensions.

3 Process of birth, death and life between lives for conscious entities.

4 Spiritual abilities conscious entities acquire in different dimensions.

Section 1: The Formation of Earth's Dimensions

Occult Chemistry contains the bulk of the information referenced here. All the dimensions in the diagram—from the physical to the causal—are centered around earth. *Occult Chemistry* describes the space in which earth exists as being carved out of the surrounding formless dimension we call space in much the same way that a bubble of air pushes water around it as it rises to the surface. *Occult Chemistry* also says that whenever a space is made, anu immediately appear to fill the void.[76] Two principles contained in this chart component are:

1 How the dimensions were created, and

2 How their sizes relate to each other.

I speculate that each of earth's dimensions have been created by quantum disentanglement, which is discussed in chapter three. This speculation is plausible because it can explain why no antimatter exists in the physical dimension when two universes separate

(Figure 10). This is the result of disentanglement--the disentangled dimension's matter is "anti-matter," or oppositely-charged matter, with respect to the dimension from which it disentangled. The physics formula for standing wave vibration on a string (page 52) explains why the ratio between tension and density must remain the same across multiple dimensions (so the frequency range within each dimension remains the same). It is important to maintain this continuity, given that these dimensions interpenetrate (see Figure 9) and co-exist with one another.

One can also argue that these interpenetrated, co-existing, wave-based dimensions explain the Many Worlds Interpretation of Quantum Mechanics and that the strings that compose anu explain the String Theory prediction that the smallest substance of matter is composed of strings. Additionally, chapter eight proposes that four spacial dimensions exist in our reality—not three spacial dimensions plus time that current scientific theory recognizes. This proposal also justifies the existence of infinity in nature, exemplified by the dimension of scale.

Changing the size of one dimension relative to another explains how each dimension can have a different density (Figure 14), which is mentioned in Theosophical literature.[77] This concept comes into play in the earthly birth and death of conscious entities in chapter seven—wherein death is explained as the shedding of our physical body when our consciousness moves from the denser physical dimension to a less dense non-physical dimension. Theosophical literature listed at the end of chapter seven notes that non-physical bodies are also shed when conscious entities move from one non-physical dimension to another.

In Figure 30, each of earth's dimensions is created as a result of its quantum disentanglement with the dimension immediately above it, as discussed in chapter three's further reading. I also speculate that the creation of anti-matter is the catalyst that actually enables the disentanglement from matter to occur. This is why, in the chart, every two lower dimensions created from a single higher one has opposite charges, as denoted by the positive and negative symbols in each dimension's circle.

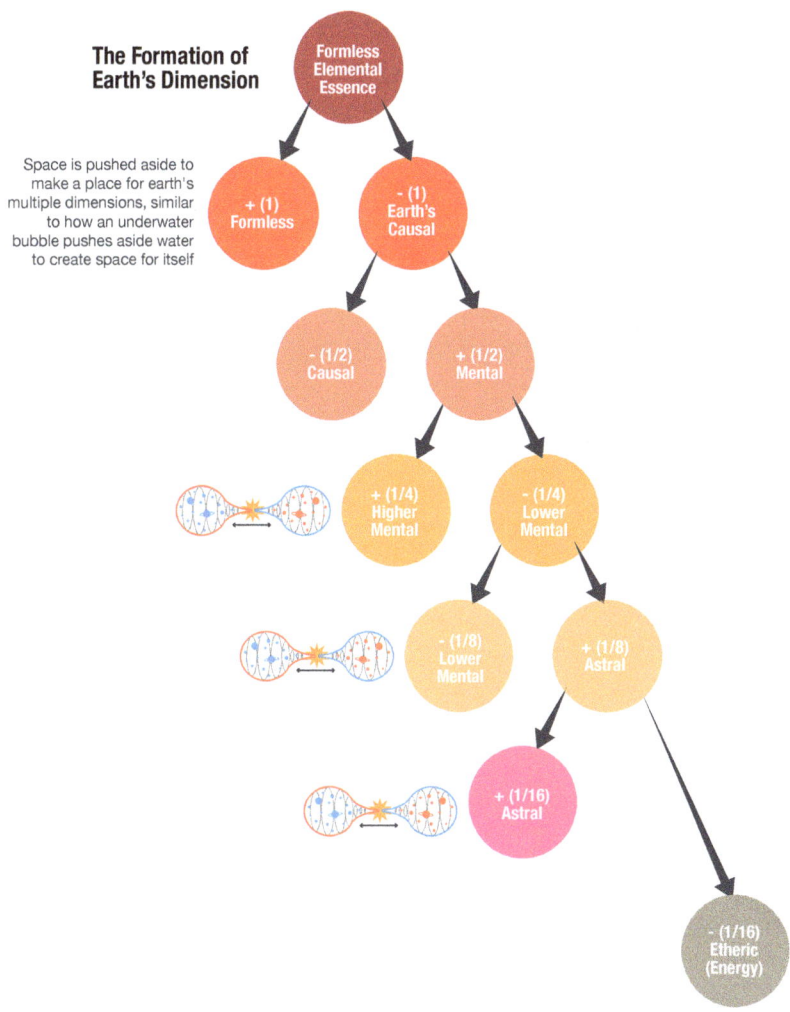

Figure 30: The formation of earth's non-physical dimensions

To align with the changing densities stated in Figure 14, each lower dimension should be denser than the one above it, a relationship established as the lower dimension is created, as illustrated in Figure 30. The methodology here, in creating two dimensions from one, suggests that the two disentangled, and oppositely charged, dimensions become half the parent's size. This creative process simultaneously compresses the matter within each new dimension, making them denser and increasing the tension within them, thus preserving the ratio of density to tension described in the formula in chapter four's further reading.

To summarize how earth's different dimensions could have been created:

- Space for earth's causal dimension was "carved out" of space's formless dimension.
- Each succeeding dimension was created through quantum disentanglement.
- The size of the two new dimensions is half that of the original, which maintains the ratio of density to tension in the physics formula for the standing wave vibration of a string.

Section 2: Balancing of Energy Between Dimensions

Figure 31 picks up with the dimensions created through quantum disentanglement, with the positive and negative charges in each circle the result of disentangling dimensions of matter and anti-matter. Here, I speculate that energy needs to be balanced between dimensions to account for imbalances that would occur when large numbers of conscious entities move between earth and its non-physical dimensions.

Using quantum disentanglement to create alternating charges between dimensions enables the generation of energy ($-\!\!\!\!/\!\!\!\!\!\sqrt{}-$) in each dimension through controlled matter / anti-matter reactions inside anu ( ), as discussed in chapter three's further reading. This speculation is beneficial for two reasons:

1 It provides energy to support celestial motion needs in each dimension, and

2 Matter / anti-matter reactions could produce small bursts of light energy within anu,[78] the cumulative effect of which is the auras seen around life forms in our dimension.

To summarize how anu could balance energy levels between dimensions:

- Anu could create energy between dimensions through controlled matter / anti-matter reactions.

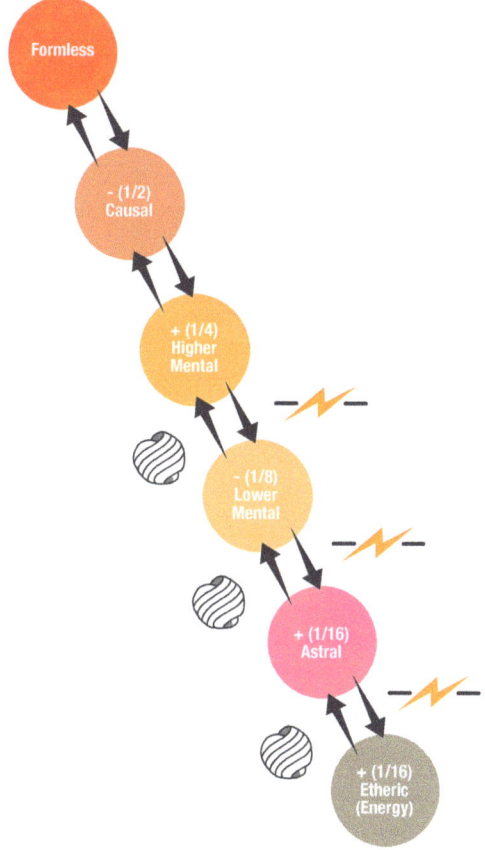

Maintaining an Energy Balance Across Dimensions

Formless

- (1/2) Causal

+ (1/4) Higher Mental

- (1/8) Lower Mental

+ (1/16) Astral

+ (1/16) Etheric (Energy)

Figure 31: How dimensions maintain an energy balance

- The generation of energy helps maintain a balance between dimensions as conscious entities move in large numbers between them.
- The controlled generation of energy between dimensions could help power celestial motion as well as create auras seen around living things.

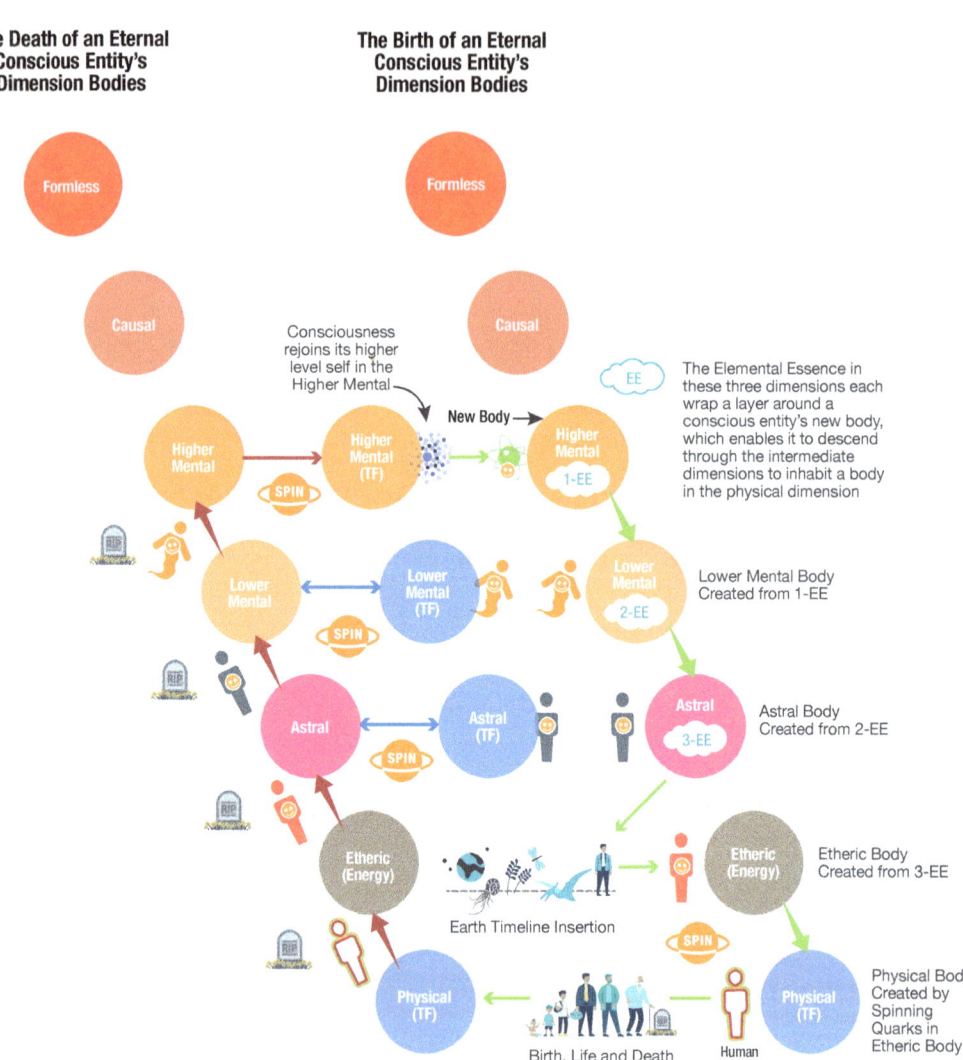

The Death of an Eternal Conscious Entity's Dimension Bodies

The Birth of an Eternal Conscious Entity's Dimension Bodies

Formless

Formless

Causal

Causal

Consciousness rejoins its higher level self in the Higher Mental

New Body →

EE

The Elemental Essence in these three dimensions each wrap a layer around a conscious entity's new body, which enables it to descend through the intermediate dimensions to inhabit a body in the physical dimension

Higher Mental

Higher Mental (TF)

Higher Mental 1-EE

SPIN

Lower Mental

Lower Mental (TF)

Lower Mental 2-EE

Lower Mental Body Created from 1-EE

SPIN

Astral

Astral (TF)

Astral 3-EE

Astral Body Created from 2-EE

SPIN

Etheric (Energy)

Etheric (Energy)

Etheric Body Created from 3-EE

Earth Timeline Insertion

Physical (TF)

Physical (TF)

Physical Body Created by Spinning Quarks in Etheric Body

SPIN

Birth, Life and Death

Human Aura Bodies

Figure 32: The process of birth and death for an eternal consciousness

Section 3: The Process of Birth, Death and Life Between Lives for Conscious Entities

Figure 32 is a visual representation of the wave-based explanation of death found in chapter seven. I do not repeat its justifications in this chapter but do include content from other chapters. Additionally, Theosophical literature states that the higher and lower mental and astral dimensions are inhabited,[79] so if readers want to find out more about these beings, they can read it directly from the source. Now for a quick review of the icons in the chart.

Conscious entities complete multiple iterations of the rectangular path outlined in this chart. The green descending arrows on the chart's right side are events that occur before one's birth in the physical dimension. The horizontal green line at the bottom of the chart represents one's life on earth. The red ascending arrows on the chart's left side describe events that occur after one's earthly life has ended. The horizontal line at the top of the rectangle marks the end of an entity's latest earthly existence, as well as the beginning of its next "life."

The non-physical bodies in each dimension have different characteristics ( ) because each dimension varies in its density. The clouds (EE) with EE on them represent masses of elemental essence (EE). When a conscious entity in a higher dimension descends into the physical dimension, it does so by adding a layer of elemental essence as it goes through each intermediary dimension, as described in Figure 9. The list at the end of chapter seven references the "death," or shedding of bodies (signified by the tombstone icon) in each non-physical dimension. It shows that what we call "death" occurs in every form-based dimension.

The two blue circles in Figure 32 represent thoughtform-based astral and mental worlds in which conscious entities live between their lives on earth (see chapter eleven). These worlds appear similar to our physical (blue-circled) existence here on earth because the spin of quarks in non-physical bodies in those dimensions (SPIN) create angular momentum, which in turn creates mass, as discussed in chapter two's further reading.

The Process of Birth and Death

We begin a quick loop around the birth and death rectangle, starting at the "New Body" arrow in its upper-right-hand corner. A single conscious entity, in the form of a small spherically-shaped spiritual being, separates itself from the multiple spheres circulating around its higher-level self. The entity acquires its first layer of form (1-EE) while still in the spirit-based higher mental dimension, enabling it to descend to the form-based lower mental dimension. There, it acquires a second layer of form (2-EE) so it can descend to the astral dimension, where a third layer of form (3-EE) is applied before it descends into the etheric dimension.

Since the physical dimension is time-based, the conscious entity enters the etheric dimension at a specific point in earth's timeline. Once there, and in its etheric body, it enters a physical body—normally the body of an infant before it is born. Note that the colorful aura surrounding the physical body could result from different levels of energy ignited through controlled matter / anti-matter reactions within anu, as described in this chapter's section two.

Once born, the conscious entity lives its earthly life as depicted at the bottom of the rectangle: growing, maturing and then dying. When the conscious entity dies, it sheds its physical and etheric bodies before moving into higher dimensions. Depending on its spiritual development—or, more specifically, the frequency at which it vibrates—the conscious entity will live in one of the multiple levels in either the astral or the mental dimension's (blue) Thoughtform World. These worlds are created by, and have, mass, which they accrued from the spinning of up- and down-quarks—just as in the physical world.

At the end of its stay in the Thoughtform World, the conscious entity sheds its remaining bodies before returning to the cluster surrounding its higher-level self. There it will determine what it will learn in its next earthly life when it begins the process all over again. The process of birth and death occur entirely within the five lower levels of the earth dimensions.

To summarize, and remembering this is a conjecture, these points explain how a conscious entity moves between dimensions during the process of birth and death:
■ The entity takes on intermediate forms through the application of multiple layers of elemental essence as it descends through

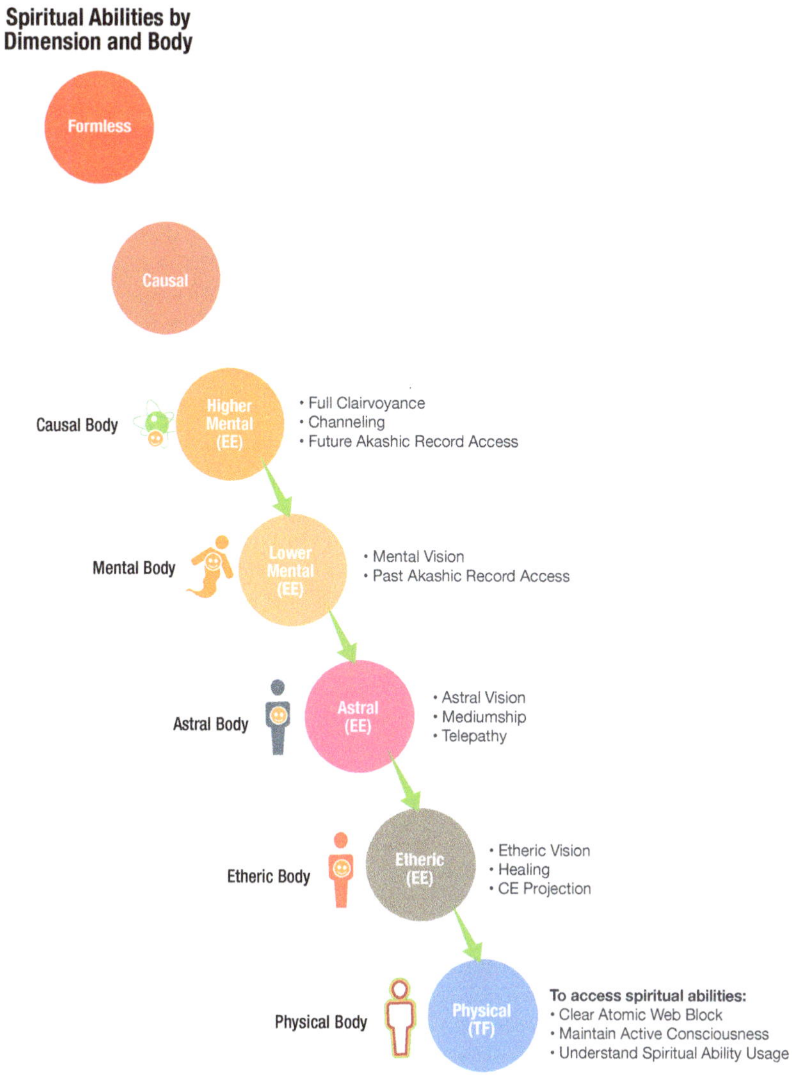

Spiritual Abilities by Dimension and Body

Formless

Causal

Causal Body — Higher Mental (EE)
- Full Clairvoyance
- Channeling
- Future Akashic Record Access

Mental Body — Lower Mental (EE)
- Mental Vision
- Past Akashic Record Access

Astral Body — Astral (EE)
- Astral Vision
- Mediumship
- Telepathy

Etheric Body — Etheric (EE)
- Etheric Vision
- Healing
- CE Projection

Physical Body — Physical (TF)

To access spiritual abilities:
- Clear Atomic Web Block
- Maintain Active Consciousness
- Understand Spiritual Ability Usage

Figure 33: Spiritual Abilities by Dimension and Non-Physical Body

intermediate dimensions to reach the desired time period in the etheric dimension.

- When the entity "dies," it sheds its denser body as it moves from a lower, more dense dimension into a higher, less dense dimension. This process happens each time it ascends into a new dimension.

- When a conscious entity resides in a dimension, it does so in the thoughtform world associated with it. When it enters a body in the thoughtform world, spinning anu cause up- and down-quarks in its non-physical body to create the semblance of mass through the cumulative angular momentum of the spinning quarks.

Section 4: Spiritual Abilities Conscious Entities Acquire in Different Dimensions

What people on earth call paranormal abilities I call spiritual abilities (Figure 33). Conscious entities in the different non-physical dimensions use these abilities to travel, communicate and manipulate objects within and between those dimensions. Humans can acquire spiritual abilities because they have bodies from the higher dimensions wrapped around their physical body (evident in its aura). Details about these abilities are described in chapter twelve, so I will not repeat them here.

Further Reading – Books on the Spiritual Development Process

Very few books discuss the mechanics of the spiritual development process, but they all describe our lives as spiritual "superhumans" after we complete our human experience—something to look forward to. One book, *The Solar System*, states that each planet in our solar system plays a role in our spiritual development, with the next stop for conscious entities after earth being Mercury's etheric dimension! The books are listed in the order of easiest read to the most difficult:

- *After We Die, What Then?* by George W. Meek
- *The Science of Spirituality* by Lee Bladon (www.evolvingsouls.com) (British)
- *The Philosopher's Stone* by Henry T. Laurency (translation of the Swedish edition)
- *The Solar System* by Arthur Powell (American Theosophist)

Henry Laurency's book can be ordered through The Henry T. Laurency Publishing Foundation in Sweden (www.laurency.com).

CHAPTER FIFTEEN
Recapping Claims, Concepts and Ideas

I wanted to take this opportunity to recap the many terms, concepts and ideas that have been explained in this book. The list is broad in scope but not overburdened by detail. This was a conscious choice because I wanted to leave it to the reader to discover their own area(s) to explore. What would be *my* next step had I been given this information? I would push for experimentation, namely the Neutron Decay Time experiment outlined in appendix A, to definitively show whether our existence is governed by particle- or wave-based physics. People will be reluctant to consider what this book presents unless there are experimental results that point to the possibility that it could be true.

Are the claims, concepts and ideas listed below all correct? I will not say so, though they may be. The objective is to research the topics, perform experiments and find out. What I will say is that the theory of wave-based physics, the interpenetration of multiple wave-based dimensions, and the claims, concepts and ideas presented in this book are internally consistent. Particle physics would be considered a special case (standing waves) in a wave-based physics paradigm, which would not diminish any of the findings in particle physics to date.

This list of claims, concepts and ideas described in this book includes (in chapter order):

Chapter Two
- Mass can be explained by the cumulative effect of angular momentum from spinning quarks in atoms.
- The technique of stopping spinning quarks in an object can explain how large stones were moved to build the Egyptian pyramids and other large stone installations (like Edward Leedskalnin's astonishing Coral Castle in Homestead, Florida).

Chapter Three

- Earth's multi-dimensional creation could have resulted from quantum disentanglement, creating the etheric and astral dimensions, with the "creation" occurring at the moment the two dimensions split into matter and anti-matter components—a cosmic form of mitosis.
- Anu-based movement of energy between dimensions could be the source of energy powering celestial and subatomic object movement in the physical dimension.

Chapter Four

- A wave-based Theory of Everything (ToE) was published in 2008.
- The strong and weak atomic forces can be explained by spinning quarks in protons and neutrons. Quark spin can be explained by the spinning and pulsing of anu within quarks as they move energy between the etheric and astral dimensions.
- Natural magnetism can be explained by anu moving non-physical world energy within magnetized iron, which can also explain magnetic field lines.
- Dark matter and dark energy could be matter and energy in non-physical dimensions that co-exist with us in the same physical space on earth.

Chapter Five

- Parapsychology can explain miracles in the Bible. This book suggests their congruence with wave-based physics.
- Quantum Mechanics' Double-Slit experiment can be explained by multiple realities, with etheric dimension-based photons simultaneously visible in multiple earth realities until an observation in one specific reality ties them to its particle-based existence.

Chapter Six

- UFO appearance and disappearance can be explained by the craft vibrating in and out of our specific reality on earth.
- UFO speed and maneuverability can be explained by the control of angular momentum and mass through quark spin described in chapter two.

Chapter Seven

■ Bodily death is the result of an active consciousness shedding its more-dense physical and etheric bodies when it moves into the less-dense astral dimension. "Bodily death" occurs in other dimensions as well.

Chapter Eight

■ Black holes can be explained as the physical dimension having four spacial dimensions, with black and white holes being the transition point between macro- and micro-scale objects.

■ The existence of infinity in nature can be explained by the existence of multiple scale orders of magnitude in four spacial dimensions.

Chapter Eleven

■ When someone thinks, their thoughts vibrate in waves around them like a radio broadcast. In higher dimensions, less developed conscious entities attempt to manifest those thoughts, with varying success based on the strength of focus (i.e., concentration), visualization and intention of the thought they acquire, explaining the comment that "thoughts are things."

Chapter Twelve

■ There are five different types of healing—each affecting different parts of a person's etheric and physical body.

Chapter Thirteen

■ It may be possible to create machines that are powered by etheric energy.

Finally, irrespective of your opinion about this book's contents, consider this point. If you hear an explanation that sounds complex, I hope it will remind you of the complexity associated with Figure 37's sixteenth-century image of planets in geocentric orbits around the earth. Maybe our model of reality is correct, but based on the wrong premise. The majority of observations in this book are multi-dimensional in scope, yet the explanations are fairly simple. Contrast this with some of the explanations you hear about subatomic and cosmic worlds from a particle / materialistic perspective.

Further Reading – Your Conscious, Subconscious and Higher-Level Self

I understand there are a multitude of opinions as to what consciousness is and what it comprises. My opinion is based on personal experiences, supplemented with Theosophical and other references.

I believe that every human being has a team of at least two other non-physical entities working toward the advancement of their conscious self. These are your subconscious and your higher-level self, or superconscious. Your conscious self is the entity actively living through your body. It makes choices that have benefits and consequences in your daily life. Your conscious self controls the voluntary systems in your body, such as your senses and your muscles. Think of it as the pilot of the commercial airliner that is your body.

Your subconscious is your body's co-pilot. It is a lesser developed human-level spirit, having more recently (than you) ascended into the human realm from animal consciousness. It is tasked with maintaining your body's involuntary systems, including its breathing, blood circulation and digestion. It does not have a language center, so it cannot speak to you, but you can speak to it and communicate with it via telepathy. It controls your emotions and drives, and as such, it is the portion of "yourself" that you must learn to control during your lifetime.

Your higher-level self (HLS) resides in the higher mental dimension. Your consciousness is a piece of this much larger self and your living of your life helps your HLS evolve. As an analogy, think of a tree. Your HLS is the tree itself—the trunk and its roots.

Every spring it sprouts a multitude of tiny leaves on its limbs. These leaves include your own conscious self and the conscious self for a number of other humans. Just as a leaf feeds the tree, your conscious self is feeding your HLS—experiences. Similarly, just as the leaves appear all over the tree's limbs, the multitude of conscious selves that come to earth can be placed anywhere in the planet's time stream to live their lives. Additionally, just as leaves die and fall away from the tree in the autumn and winter, so do the human bodies of the conscious selves when the conscious selves return to the HLS at the end of their earthly lives. The objective of each conscious self is to become its own spiritual entity. Remember—the focus of your life is the spiritual development of your conscious self, not the earthly accumulation of money, power, influence or the achievement of celebrity or social status.

Why do I believe these things?

Regarding the subconscious, Max Freedom Long's book *The Secret Science at Work* dedicates a couple chapters to teaching how to work with your subconscious. I have asked my subconscious self on multiple occasions to heal parts of my body, and moments later felt it sending vibrations to that particular area (note: I can feel it because I have sensitized my body to feel its energy). There was one night in particular when my pineal gland had been damaged by a spirit and I asked my subconscious to restore it. What it did,

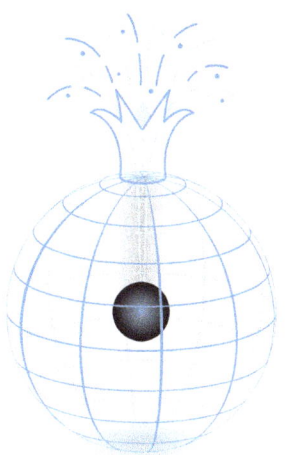

Figure 34: A causal entity

surprisingly, was give me a "fully functioning" pineal gland that roared like the engine of a Formula One race car when I vibrated it. This pleasant surprise was short-lived, however, as a spiritual entity quickly came and disabled it.

On the HLS side, the spiritual entity depicted on the cover of this book, I believe, resembles my HLS and the small balls moving around its central sphere are the conscious selves for which my HLS is responsible. Until recently, I could clearly see in my mind's eye the central sphere illustrated on the cover, from the perspective of a smaller ball situated closest to it. In December 2021, my HLS appeared in my bedroom, and a little while later, the entity in Figure 34[80] appeared next to my bed, wrapped in a semi-transparent milky sphere with mist gently spewing out the top.[81] The object's shape reminded me of a pomegranate.

Interestingly, the Hebrew Bible states that 200 metal pomegranates surrounded the tops of the two pillars of the first temple in Jerusalem and that the hem at the bottom of the chief priest's outer robe should contain alternating multicolored pomegranates and small bells.[82] This made me wonder if the symbol of the pomegranate in Judaism was a reference to a causal body. The Theosophical book *The Causal Body and the Ego* devotes multiple chapters to the development and individualization of the causal body, though there are four pages in particular that cover the initial individualization process.[83] Its description of the individualized causal body is very close to what I describe above, except the body I saw contained a mature central sphere; the initially formed causal body does not.

Conclusion – The Future Will be Determined By Our Love for One Another

I have been blessed to have had the opportunity to think about spiritual questions and the non-physical world for an extended period of time, and the answers shared here are offered to save you time in the development of your own spiritual understanding. My understanding comes down to these two beliefs:

1 A wave-based, multi-dimensional reality exists—on earth as well as in the universe.
2 For humanity to move forward into this new reality, it must conquer its fear—fear that we live a temporary mortal existence, fear of the unknown and fear of what this new future holds.

The idea that our reality may be different from what we generally believe will create great opportunity for some but will instill great fear and anxiety in many others. Over the past century, large masses of people have been controlled or silenced through fear, including the ultimate fear: the fear of death. But how will people's thinking change when they realize that they are, in fact, eternal beings—that their bodies might be killed but their consciousness will live on? Once this realization takes hold, it will be difficult, if not impossible, to rule anyone by fear ever again. How will this world be governed then, if not through fear and political intimidation? In this more ideal world, people will be governed by their love for one another. It is the only way to get eternal beings to work together!

You have just been given the greatest "otherworldly" revelation—that eternal beings living in the non-physical dimensions, whom we have yet to meet, are governed by love, and that they are waiting for us to elevate our love for one another to the point where we can join them.

This, I believe, is the greatest upcoming challenge for humanity—not to learn love on an individual basis but learn to love on a global scale—to find common ground with different people groups around the world through which we can work together. It will require a time of global peace, during which people can focus on humanity's needs.

To accomplish this objective, humans will need to work together toward a mutual goal that is bigger than themselves, a goal that will require selflessness on everyone's part to achieve a benefit. People will have to cast aside their selfish ambition and the desire to accumulate wealth, power and influence. Can it be done? Yes, but only if we can foster a genuine love for one another—all races, all cultures, all languages and all social and economic backgrounds.

In closing, it is my hope that reading this book has helped you more fully understand the possibilities for your spiritual journey here on earth. There is so much to know and discover! I hope you enjoy the life you have been given and that you work in love to help others enjoy their lives as well. Life itself is a prize to be coveted and love is the portion of that prize we get to pass along to others.

Peace.

Where to Find Additional Information

When something significant happens to people that they do not understand, whether a spiritual experience, a medical diagnosis or otherwise, they can become voracious readers on the subject to understand it. It certainly was true for me! In the late 2000s, my desire for books on spiritual topics surpassed what had been my passion for reading up to that time: books about business and technology—as well as textbooks! When I lived in Manhattan, I enjoyed perusing the textbook section of a bookstore at 16th and Broadway—fond memories. My initial spiritual interests were wide-ranging: monasticism, mysticism, philosophy, quantum mechanics, metaphysics and secular humanism.

An Interesting Technique:

When beginning my search for spiritual books, I didn't know at first which ones might interest me, so I often skimmed through the endnotes of each new book I read. If I found something interesting, I would read the content associated with that endnote. If I found that interesting, I ordered the book. It only took 3–4 iterations of using this technique to find the important works in each field of interest. If you want to read more about this book's topic, I have marked the books I value the most with an asterisk "*" in the Bibliography.

The best resources I found about exploring the non-physical world are from the Theosophical Society, and the best of these books were about a century old. I encourage you to see what they offer.

Other sources of spiritual information I found included:

- Watching virtual lecture videos; these videos taught me a great deal about conventional science, quantum mechanics, theology and philosophy.
- Collecting catalogs of spiritual book publishers or reading books from the libraries of metaphysical organizations.

- Visiting metaphysical organizations or their local chapters.
- Attending local meetup.com groups, which enable you to meet other people with similar interests and learn from their experiences.

After you become familiar with spiritual literature, you may begin to perceive how seemingly unrelated things fit together. If so, I strongly urge that you record your thoughts in a journal. Writing about your experiences helps build your understanding, one complementary thought at a time. Before I started writing books, I kept a journal in which to capture insights as they came to me. Most of this book's concepts originated in this way.

APPENDICES

If you want to dig deeper into the premise of this book and the material that supports it, please read the appendices that follow:

- Appendix A—Observations and Experiments to Falsify Claims Made in this Book
- Appendix B—Wave-Based Atomic Forces
- Appendix C—A Hierarchical Torus Structure of the Universe
- Appendix D—Framework for Spiritual Development
- Appendix E—Wave-Based Explanations for Miracles in Bible Scripture

APPENDIX A
Observations and Experiments to Falsify Claims Made in this Book

"It has turned out that a wave-only universe is the simple answer to half a century of puzzles and confusion."

—Milo Wolff, PhD, author of *Schrödinger's Universe* [84]

Observations Supporting the Claim that Physics is Wave-Based

1 **Wave-Based Quantum Energy Transfer** – John Cramer, PhD, proposed the transactional interpretation of quantum theory in his book *The Quantum Handshake: Entanglement, Nonlocality and Transactions*. It describes how energy passes between wave-based atoms using a process that resembles how computers transact their business on a network. His work depicts atoms as a combination of inward and outward waves,[85] which supports the Wave Structure of Matter (WSM) as a valid interpretation of quantum theory. Energy transfer is an important component for justifying a wave-based quantum theory because particle-based theories have not done so. Particle physics says, for example, that a photon is the carrier of electromagnetic energies between particles, but they do not explain the process or define how energy is transferred between them.

2 **Wave-Based Electromagnetics** – Carver Mead, PhD, corrected known flaws in Maxwell's Equations' magnetic terms using data based on the WSM. His book *Collective Electrodynamics* is widely used in Silicon Valley to solve electromagnetism problems with transistor circuits. This is important because particle-based physics could not rectify the equation's flaws—another justification for the Wave Structure of Matter.

"The quantum world is a world of waves, not particles. So we have to think of electron waves and proton

waves and so on. Matter is "incoherent" when all its waves have a different wavelength, implying a different momentum. On the other hand, if you take a pure quantum system—the electrons in a superconducting magnet, or the atoms in a laser—they are all in phase with one another, and they demonstrate the wave nature of matter on a large scale. Then you can see quite visibly what matter is down at its heart."

—Carver Mead PhD, *author of Collective Electrodynamics* [86]

3 Theory of Everything – Particle physicists have labored for decades to develop a set of mathematical equations that explains all phenomena in the physical universe, but this Theory of Everything (ToE) has been elusive. However, a wave-based Theory of Everything was proposed in 2008 by Milo Wolff, PhD, who published a three-equation ToE in *Schrödinger's Universe* [87] that makes the case for a fully wave-based reality. Wolff also expressed amazement about how these equations aligned with other laws of nature. His three ToE equations are summarized in Figure 35.

- Quantum matter waves exist in space and are solutions of a scalar wave equation where Φ is a scalar amplitude, *c* is the velocity of light, and *t* is time.

$$\nabla^2 \Phi - \left(\frac{1}{c^2}\right)\partial^2\Phi/\partial t^2$$

- Waves from all particles in the universe combine their intensities to form the wave-medium density (space) at each point in the universe, according to Mach's Principle. In addition, *m* is mass at rest, *h* is Planck's Constant, *f* is frequency, and *r* is radius. It states that the mass (*m*) or frequency (*f*) of a particle depends on the sum of the squares of all wave amplitudes from the particles inside the observable universe (approximately 10^{80}).

$$Space\ density \propto mc^2 = hf \propto \sum_{1}^{n}[\frac{\Phi_n}{r_n}]^2$$

- The total amplitude of particle waves at every point always seeks a minimum, according to the minimum amplitude principle (MAP).

$$\sum_{1}^{n} \Phi_n = a\ minimum$$

Figure 35: Milo Wolff's wave-based Theory of Everything, reproduced from Schrödinger's Universe

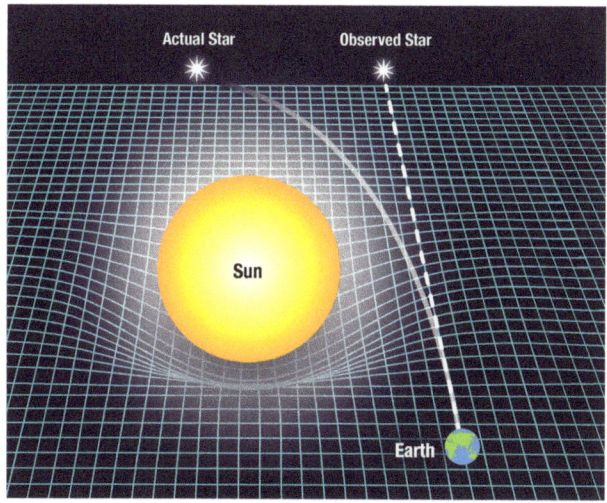

Figure 36: The bending of light in Einstein's Theory of Relativity

4 Einstein's Theory of Relativity – When scientists speak about the bending of light in Einstein's Theory of Relativity, they claim it's due to gravity, but illustrations that describe it (Figure 36) show a two-dimensional, flat "fabric of space" along which the light is bent, not a three-dimensional space. Given Milo Wolff's Theory of Everything, the fabric of space may be more accurately represented as a three-dimensional sea of waves. The depressions around celestial objects in this three-dimensional sea could be seen more as bubbles than depressions, and just as bubbles displace the water that surrounds them, the bending of light around the sun and other celestial objects could be similar to the diffraction of light around bubbles in water—instead of gravity.

How to Falsify the Claim That Physics is Wave-Based—Measure Neutron Decay Time in Space

A February 13, 2018, *Quanta Magazine* article[88] documented an approximately nine-second discrepancy in neutron decay times between "bottle" (about 879 seconds) and "beam" (about 888 seconds) techniques, the two primary methods to measure neutron decay. Scientists have been unable to explain this discrepancy. One possible answer to this conundrum is that matter is wave-based, not particle-based.

In *Schrödinger's Universe*,[89] Milo Wolff, PhD, proposed an experiment to verify Wave Structure of Matter (WSM) energy transfer by measuring discrepancies in neutron decay time. The experiment hypothesizes that WSM decay times will vary depending on the coupling between two vibrating wave-like entities—and also will depend on the distance between them. Since the experiment calls for measuring neutron decay times in outer space, Wolff theorized that neutron decay times in space should be similar to the decay times on earth if neutrons were particle-based, because neutron decay is an intrinsic property of the particle. However, if neutron decay times in space are significantly different from times observed on earth, it could indicate that neutrons are wave-based, which would align with the WSM model.

Two space-based neutron decay experiments have already been undertaken on NASA missions. The missions were: 1) The MESSENGER flybys of Mercury and Venus (about 780 seconds) and 2) The Lunar Prospector (about 887 seconds). Additional work needs to be done to increase rigor around the testing and accuracy of the results, but I applaud these initial attempts and the work that went into them. One thing I would ask is that future experiments include scientific consideration that variations from earth-based neutron decay mean times may indicate that physics is wave- instead of particle-based.

Observations Supporting the Claim of a Multi-Dimensional Reality

> **"The necessity of the quantum in the construction of existence: out of what deeper requirement does it arise? Behind it all is surely an idea so simple, so beautiful, so compelling that when—in a decade, a century, or a millennium—we grasp it, we will all say to each other, how could it have been otherwise? How could we have been so stupid for so long?"**
> —John Archibald Wheeler, physicist[90]

My opinion, based on having researched and written four books on the subject, is that it will be difficult to detect dimensions beyond the etheric if the experiments are conducted in the physical dimension. However, the etheric dimension should be detectable given properly designed experiments. One area to research for evidence of the existence of the etheric dimension could be to study the energy-based component of the human body.[91]

Here are three observations to support the claim that reality is multi-dimensional.

1 **Visually Seeing Healing Energy** – As suggested in chapter twelve, two noted psychics, Olga Worrall and Ingo Swann, participated in an experiment to find out if healing energy could be seen emanating from their hands.[92] The test, using a cloud chamber, showed that both participants were able to create visible movements within the chamber when healing energy passed from one hand to the other. Ms. Worrall also achieved similar movements within the same cloud chamber—when she was six hundred miles away from it!

2 **Propagation of Light in a Medium** – Physicists say that for waves to propagate, there must exist a medium through which they can propagate—like ripples through water and sound vibrations though air. Light propagation, however, appears to lack such a medium because outer space is commonly understood to be a vacuum. If reality is multi-dimensional, then maybe light does not propagate through the physical but rather through the etheric. This would make more sense as the etheric is described as a dimension of energy. Additionally, maybe we should consider whether other forms of energy—such as electricity and magnetism—are active in those dimensions as well. This is what Dr. Wolff's Theory of Everything equations seem to describe.

3 **Complexity of Quantum Mechanics's Mathematics** – Many predictions made by Quantum Mechanics, the physics of subatomic particles, have been verified experimentally with accuracies reaching multiple decimals. Despite these successes, however, challenges remain, perhaps the most enduring of

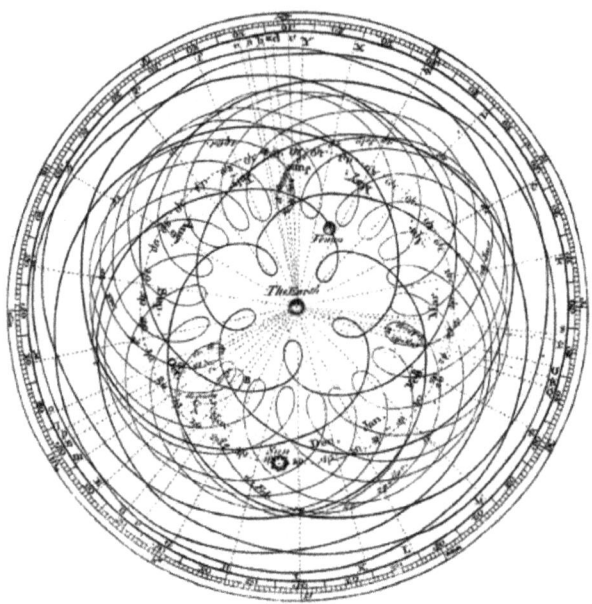

Figure 37: A geocentric (Ptolemaic) model of the solar system,
with the earth at its center (public domain illustration)

which is the complexity of the mathematics required to arrive at these predictions.

A similar situation occurred in the 1600s, an era in which most of humanity believed in geocentrism—that the earth was the center of the solar system and the other planets revolved around it.[93] Complex calculations had to be concocted to account for the multiple loops needed to make the math work for each planet's intricate orbit (Figure 37). When heliocentrism, the idea that the earth orbited around the sun, was adopted, planetary motion and the mathematics to calculate it became greatly simplified.

Could a similar "blindness to reality" now be in play in Quantum Mechanics—a twenty-first-century parallel to geocentrism!? If we read the descriptions of the etheric dimension components in the atomic elements described by Annie Besant and Charles Leadbeater late in the nineteenth century, one might conclude that Quantum Mechanics is—in reality—the physics of the multi-dimensional, not the subatomic, which fits comfortably within a wave-based physics worldview. (Note: Carver Mead PhD made a similar comment, which I found as this book was going to press. See the reference in endnote #86, page 9.)

How to Falsify the Claim That Multiple Dimensions Exist #1—Discover Anu

I've said so much about anu that it would be appropriate to see if it can be found. I've also often wondered if instruments such as the Large Hadron Collider in Switzerland could detect the anu's action within quarks—perhaps by detecting energy bursts created as anu move bits of energy from the astral into the etheric dimension (my prediction). *Occult Chemistry* describes the action of the (torus-shaped) anu, which is composed of ten loops—three thicker than the other seven—this way:

"In the three [thicker loops] flow currents of different electricities; the seven [thinner loops] vibrate in response to etheric waves of all kinds—to sound, light, heat, etc.; they show the seven colors of the spectrum; give out the seven sounds of the natural scale; respond in a variety of ways to physical vibration—flashing, singing, pulsing bodies, they move incessantly, inconceivably beautiful and brilliant."[94]

Alternatively, *Occult Chemistry* says that when an electric current is introduced (Figure 38), anu slow down and line up in parallel lines so that energy flows out of one anu and directly into the next.[95] It makes me wonder if anu could be the source of electricity and

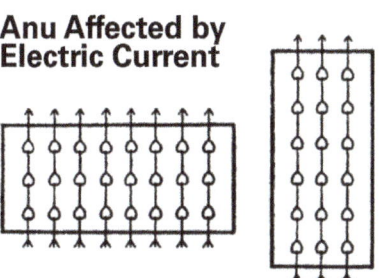

Anu Affected by Electric Current

Figure 38: *Occult Chemistry* illustration showing how anu line up when exposed to electricity.

electro-magnetism, which I mentioned in an earlier article.[96] Lastly, experiments could be done on magnets to search for anu.

How to Falsify the Claim That Multiple Dimensions Exist #2—Subatomic Particle Desk Study

Three editions of *Occult Chemistry* were published in which Theosophists Annie Besant and Charles Leadbeater used their spiritual abilities to visually examine[97] naturally occurring atomic elements. They produced over four hundred pages of content and over two hundred drawings of the physical structure of those elements, showing how they decompose into etheric dimension-based subatomic structures. The claim that these multiple dimensions exist could be falsified by performing a desk study comparing the multi-dimensional decomposition of the atomic elements listed in *Occult Chemistry* with the subatomic particles science has discovered in experiments at CERN's Large Hadron Collider (LHC) and at other particle accelerators. If there is a correlation, then the existence of multiple dimensions should be further studied. Finding no correlation, however, would not falsify the existence of multiple dimensions but would call into question the decomposition of elements depicted in *Occult Chemistry*.

For the record, there are other correlations with modern scientific findings in *Occult Chemistry*. The third edition of the book includes a "number-weight" for elements that very closely matches each element's atomic weight, although the numbers in *Occult Chemistry* were published before scientists started using mass spectrographs to measure the atomic weight and mass of atoms. As stated in *Essential Concepts*, the book's authors also described quarks within the Hydrogen atom, although they did not have a name for what they saw. (Scientists discovered quarks over fifty years later—in 1964!) Since the two theosophists identified quarks before they were "discovered" (or even named) by science, fraud or the manipulation of data can be ruled out as an explanation for their claims.

Observations Supporting the Existence of Beings in Wave-Based Dimensions

If reality is wave-based and multi-dimensional, it also is reasonable to consider whether living things, both human and non-human, exist within those wave-based dimensions. Unfortunately, it is difficult to run experiments in the physical dimension that test for life in other dimensions. However, we do have written observations from across more than two hundred years that convey the writers' experience interacting with non-physical beings. According to their works, the non-physical world is teeming with life:

"If there are intelligent beings in the physical universe, we failed to find them.... In the nonphysical universe, it was an entirely different matter. We encountered hundreds, if not thousands, most of them non-human."
—Robert Monroe, out-of-body explorer and author.
Ultimate Journey, page 13.

"We must now attempt to fill in the figures—to describe the inhabitants of the astral plane. The immense variety of these entities makes it exceedingly difficult to arrange and tabulate them. Perhaps the most convenient method will be to divide them into three great classes— the human, the non-human and the artificial."
Charles Leadbeater, Theosophical Society leader and author. *The Astral Plane*, pages 29-106. Note that these pages discuss beings in the astral dimension.

"In our endeavor to describe the inhabitants of the mental plane it will perhaps be well for us to divide them on the same three great classes chosen in the manual on the astral plane—the human, the non- human, and the artificial—though the sub-divisions will naturally be less numerous in this case than in that, since the products of man's evil passions, which bulked so largely there, can find no place here."
—Charles Leadbeater, Theosophical Society leader and author.
The Devachanic [Mental] Plane, pages 30-100. Note that these pages discuss beings in the mental dimension.

"In general, it must be maintained that all things I have written in this book have not been written in any other way than from actual experience, from conversation with spirits and angels, from thought communicated as tacit speech.... For I have in every case perceived their presence. (1748, 23 Aug.)"

—Emanuel Swedenborg, philosopher, mystic and author. ...*Recounting Spiritual Experiences during the years 1745 to 1765*, paragraph 2894. Swedenborg amassed four volumes of experiences.

How to Falsify the Claim that Beings Exist in Multiple Dimensions—Physical Mediumship

To falsify the claim that beings exist in multiple dimensions, I recommend studying the energetic form, not the protoplasmic form, of physical mediumship sessions that have been conducted over the years.

Physical mediumship, as opposed to the better-known mental mediumship, involves the five physical senses and the manipulation of energy and energy systems. Mental mediumship, in contrast, is primarily related to inter-dimensional telepathic communication. Similarly, the energetic form of physical mediumship involves the blending of energies between dimensions, which includes the participation of multiple living entities on both sides. Protoplasmic physical mediumship, in contrast, is focused on the medium leading the session in this dimension.

Robin Foy, along with his wife Sandra, have over forty years of involvement with the study of physical mediumship and its related phenomena. Their most well-known work is founding the Scole Experimental Group in Scole, England, with Diana and Alan Bennett. Over a five year period between 1993 and 1998, the Scole Group held over one thousand hours of physical mediumship and phenomena demonstrations, many of which were open to the public. Some of these demonstrations were attended by three investigators from the Society for Psychical

Research (SPR) in London. Their report, titled *The Scole Report*, was based on over thirty sittings that spanned two years.[98] A video documentary of the Scole Experiment, titled *The Afterlife Investigations*, can be viewed on the Internet.[99]

In 2021, the Foys helped start the Spiritual Science Founders Association in Antequera, Spain, (www.ssf-robin-foy.com) which promoted the scientific study, development, demonstration of and education about physical mediumship. The physical mediumship group meeting there was known as the Torcal Experimental Group.

APPENDIX B
Wave-Based Atomic Forces

Several years ago, I became curious about how protons and neutrons are bound together and found a simple way to model their atomic forces. To do this, I built atomic nuclei using several hundred 25mm magnetic hematite spheres. Clearly, I'm not a scientist (just to set expectations), but I did learn a few things while playing with them. The list below and the drawings that follow summarize my observations.

1 I noticed right away that the magnetic spheres like to form six-sphere loops. Since there are three quarks in each proton and neutron (six total in a proton-neutron pair), this led me to believe that quarks created nuclear forces, not protons and neutrons.

2 Taking the insight that quarks might be the source of nuclear forces, I configured multiple layers of magnets in the shape of different atomic elements—using three quarks for each proton and neutron. Figure 39 on the next page gives you a visual representation of what I found in the neon atom:

a. The magnetic spheres stacked up the way they are illustrated in the atom's nucleus.

b. The two rows of blue magnetic spheres in the middle of the atom created the orange X and Y electron orbital shells.

c. The row of blue spheres, on the top and bottom of the atom, created the top and bottom (respectively) Z electron orbital shell.

d. The four rows of pink magnetic spheres (two rows are behind the two rows of blue spheres in the middle of the model nucleus) create the pink and blue s-shell electron orbital shells in the center of the model nucleus.

e. A triangle of spinning up- and down-quarks (shown below the neon atom) created the conical shape of the X, Y and Z electron orbital shells.

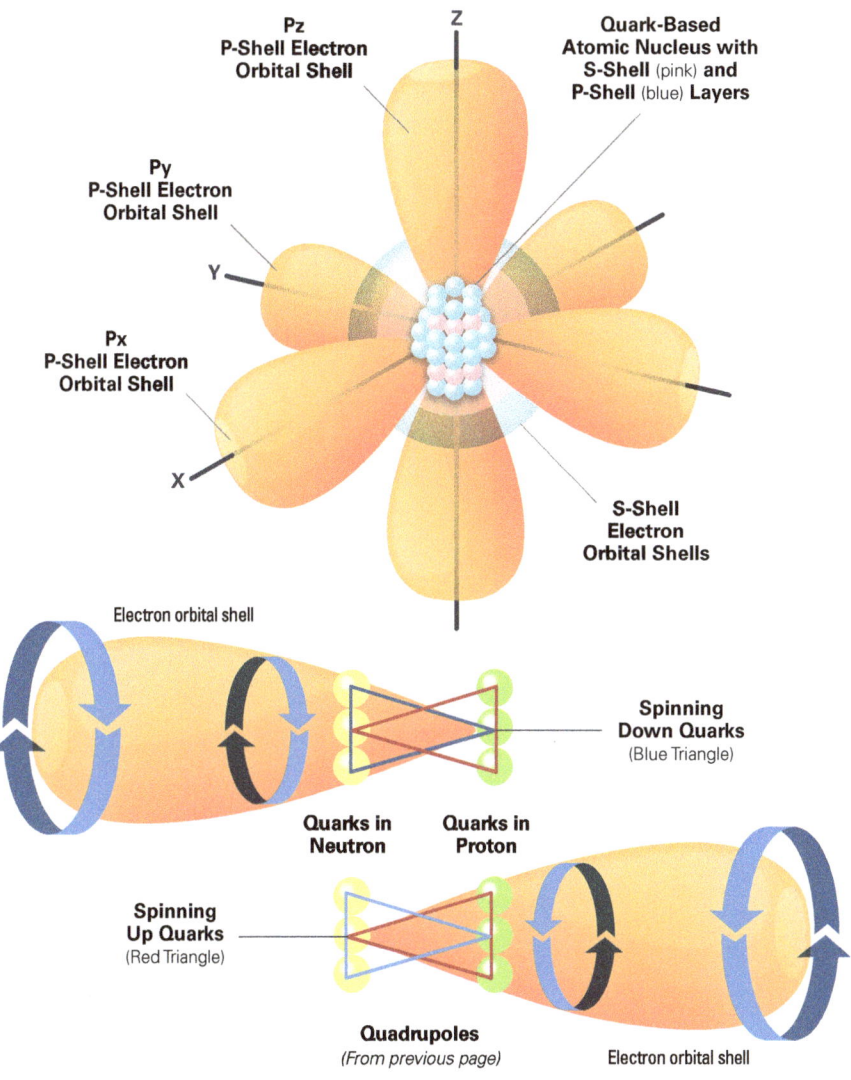

Figure 39: Neon atomic nucleus and electron orbital shells

3 When I tried to model larger atoms, their magnetic representations became more layered and complex, but surprisingly—they aligned precisely with the atom's electron orbital shells. There were two type of layers—one related to the row of the Periodic Table of Elements that the atom occupied, and the other layer to the electron orbital shell it occupied (Figure 40 and Figure 41).

Cross-Section of Quarks in an Atomic Nucleus with Electron Shell Dipoles and Quadrupoles Identified

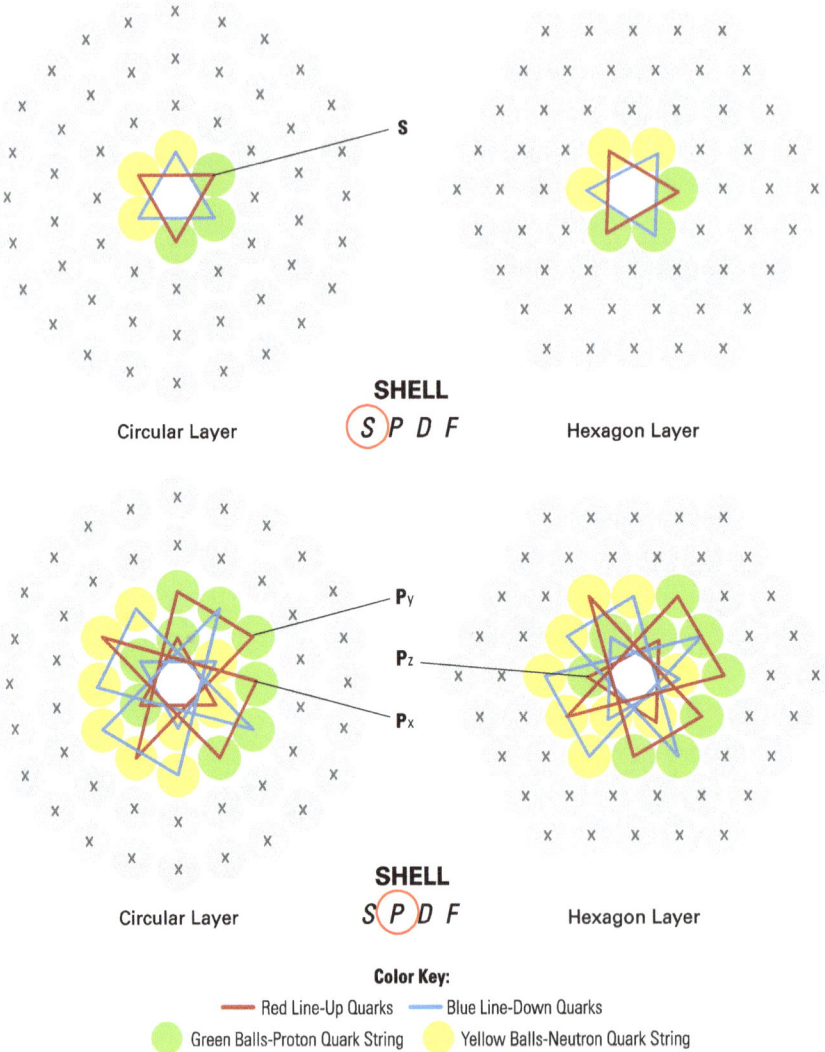

Figure 40: The dipoles and quadrupoles that create S- and P-Shell electron orbital clouds

4 In the outer electron orbit shell layers, I noticed that the magnetic spheres lined up with an offset—on their own! One type of ham radio antenna, a Yagi (beam) antenna, uses an offset in its design to produce a conical signal shape that helps operators

Note: In the F-Shell, the magnetic spheres in the Hexagon Layer assume a Circular Layer shape.

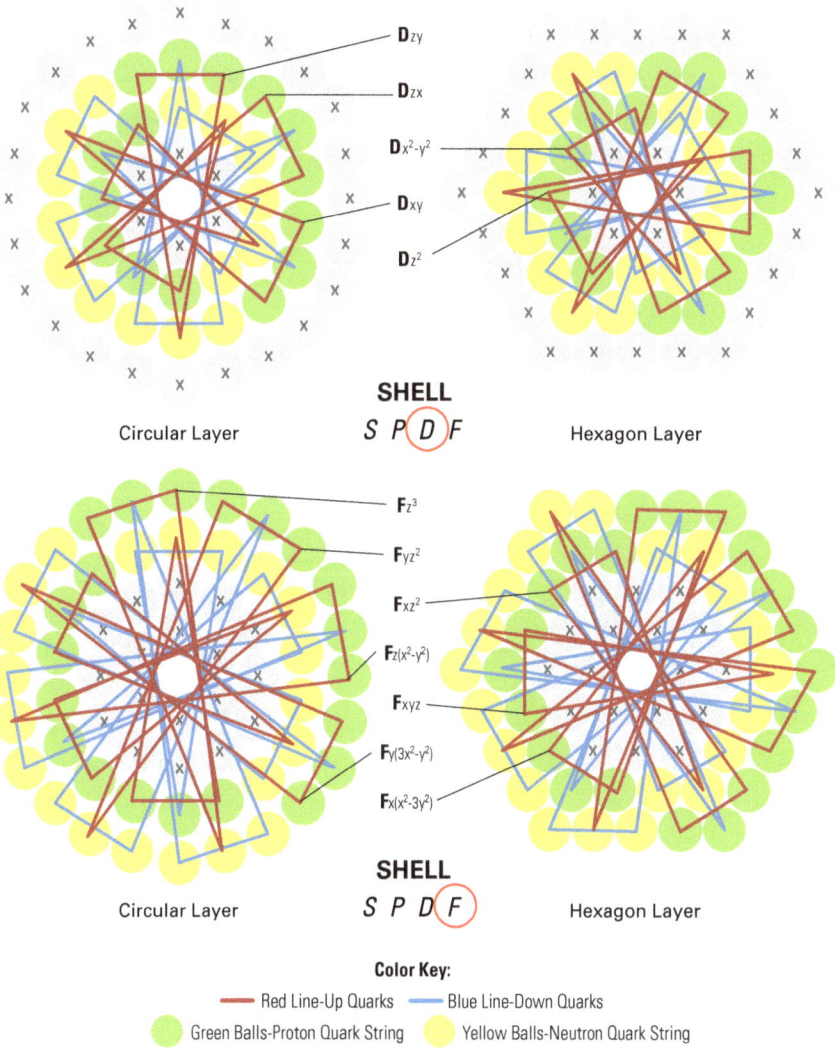

Figure 41: The quadrupoles that create D- and F-Shell electron orbital clouds

aim their transmissions in specific directions. Surprisingly, the degree of offset in the magnetic spheres matched the degree of the conical shape of that element's electron orbital shell. (In fact, I used an online simulator to help visualize the different radiation patterns.[100])

To summarize, using magnetic spheres to model atomic forces helped me understand how electron orbital shells were created. It also strongly hinted that positive and negative charges came from quarks, not protons, neutrons or electrons. This left me wondering whether electrons existed at all! If you want to explore this concept further, my third book, which is listed in the Bibliography, covers these charts in greater detail.

More Justification for Using Quarks to Model Atomic Forces

While editing this appendix, I noticed that the up- and down-quarks in the Hydrogen atom drawn in *Occult Chemistry* have the same arrangement as the up- and down-quarks I drew in Figure 39. It's exciting to find independent corroboration for your thoughts from a drawing that was created a century ago. The series of images below show how the Hydrogen atom illustration from *Occult Chemistry* aligns with this appendix's description of how electron orbital clouds are created. The last two images (Figure 45 and Figure 46[101]) compare *Occult Chemistry's* Hydrogen atom drawing with the first image ever taken of the Hydrogen atom. Note how the drawing's outer shell and the electron orbital cloud inside the shell align with structures in the image.[102] **How to invalidate this claim:** *Occult Chemistry* says that atoms of—Hydrogen, Nitrogen and Oxygen were never observed to move in pairs—except in Deuterium.[103] See if these elements can be observed in pairs. It also describes the shape of the sphere enclosing the Nitrogen atom as an ovoid, not as a circle.[104]

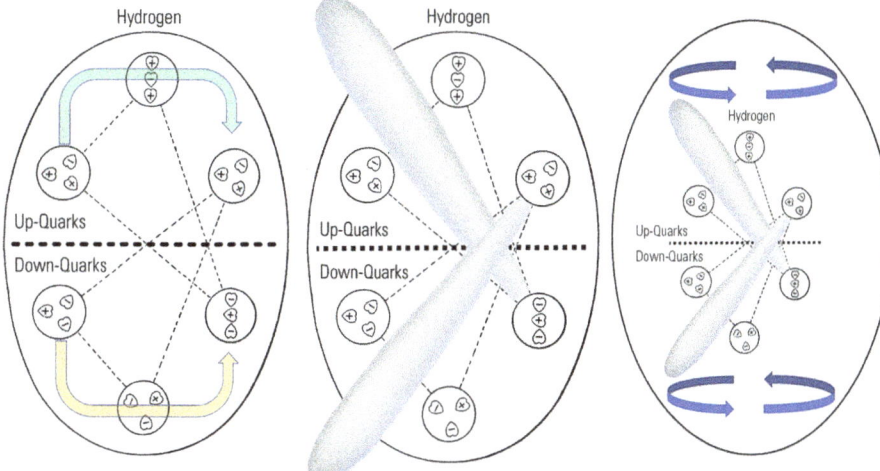

Figure 42: Separation of up- and down-quarks in the Hydrogen atom to create quadrupoles
(image from *Occult Chemistry*)

Figure 43: Hydrogen atom with the addition of quark-based electron orbital cloud

Figure 44: Moving the ovoid-shaped outer shell outward, so the orbital cloud can remain inside it

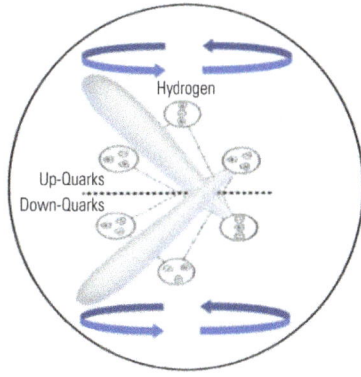

Figure 45: Changing the outer shell to a circular shape

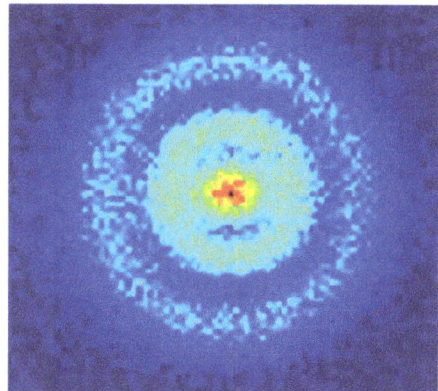

Figure 46: The first image of a Hydrogen atom (note the similarities with Figure 45)
(photo credit in Endnote)

APPENDIX C
A Hierarchical Torus
Structure of the Universe

This appendix follows up on my comments in chapter eight that 3D-printed physical models of a torus (Figure 47) visually explain how a torus can maintain its structural integrity.

The photo of the center portion of a physical torus shown in Figure 48 highlights its inner core, which consists of small circles. This made me wonder whether the structural integrity of a large torus results from its inclusion of smaller tori in its core.

These small tori appear to be the next lower scale in order of magnitude than the scale represented by the large torus. This could mean that tori of various sizes, such as the torus that surrounds a person and the torus-shaped magnetosphere that surrounds the earth, could be connected (Figure 49). The multiple tori that humans create could provide the structural integrity for a larger torus, although it may not be for the larger torus that protects our planet. If this conjecture is correct, it could

Figure 47: A 3D-printed physical representation of a donut-shaped torus

Figure 48: Small tori in the core of a 3D-printed torus

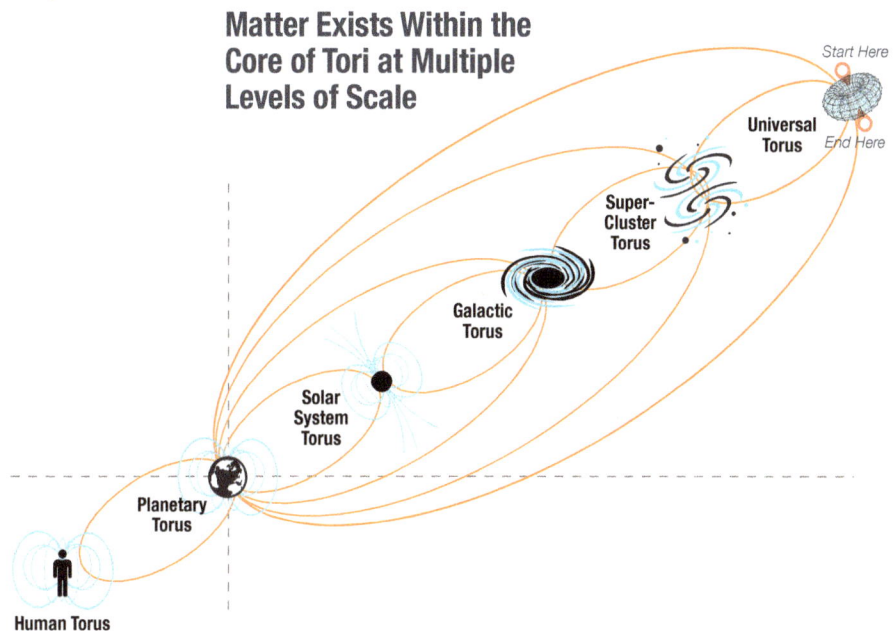

Matter Exists Within the Core of Tori at Multiple Levels of Scale

Start Here

Universal Torus *End Here*

Super-Cluster Torus

Galactic Torus

Solar System Torus

Planetary Torus

Human Torus

Figure 49: Matter exists within the core of tori at multiple levels of scale

mean that the torus shape integrates matter from the very smallest particles to the very largest structures in the physical universe. It also ties our human existence to the structure of the universe.

Now let's examine the energy torus that surrounds our planet, the magnetosphere, in greater detail. This torus-shaped shield that protects the earth from harmful solar winds and radiation is shown in Figure 50. We earlier considered whether the magnetosphere could get its structural integrity from the tori surrounding human beings living on the planet, but how? Fortunately, our physical torus model gives us a possibility. As the photo in Figure 51 shows, the same base 3D-printed structure can form two different torus shapes—the traditional donut shape and a less-familiar globe shape. Both shapes in this instance were created from the same 3D printed shape.

If you overlay these two structures shown in Figure 50's magnetosphere illustration, it looks like the globe-shaped torus resides INSIDE the donut-shaped torus. Could the integration of

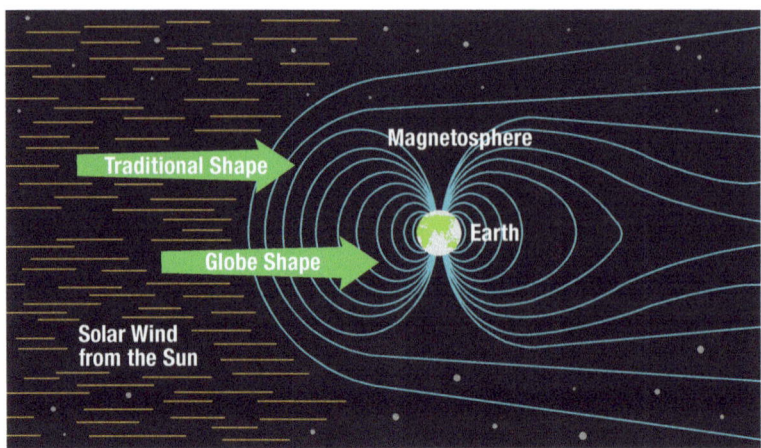

Figure 50: Two tori defined within the earth's magnetosphere

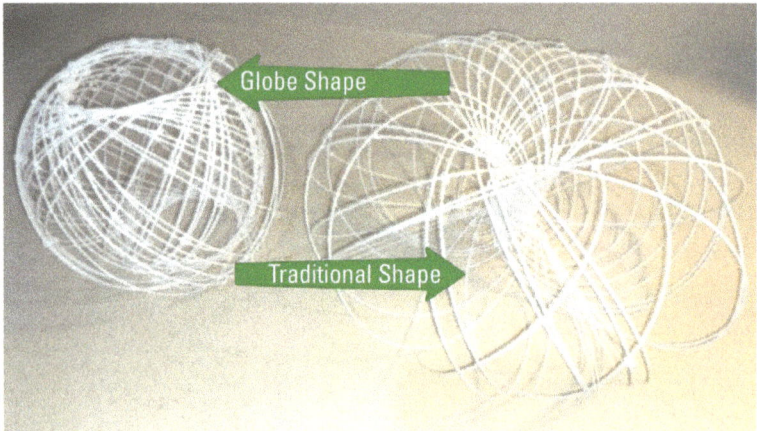

Figure 51: Two shapes of a 3D printed torus

donut- and a globe-shaped tori occur in nature? The next two figures demonstrate how they can be constructed—by looping the two tori together.

The photo in Figure 52 shows how the two 3D-printed pieces look before they are shaped. Figure 53 shows the two loops after they are shaped. One loop is shaped to create a donut-shaped torus shape while the other loop is shaped to become a globe-shaped torus that sits inside the donut-shaped torus. It's not readily apparent in the photo, but the smaller tori in the globe-shaped torus are evenly distributed around the torus, not concentrated in a central core as in the donut shape (the concentration is not as apparent in this photo because the model in Figure 52 is much smaller than the model shown in Figure 46). This implies that smaller tori associated with the globe shape in the magnetosphere could be more evenly distributed around the surface of the earth. The globe-shaped torus's structural instability means that it must rely on the larger donut-shaped torus for protection from the Sun's solar winds and radiation (see sidebar).

Figure 52: Looping two tori together

Figure 53: Two looped tori—the final shape

The double-torus structure of the magnetosphere reminds me of the human family relationship, where one partner provides protection and stability so the other partner can focus on nurturing and raising the family's children. Seeing these two intertwined tori in Figure 52 similarly made me wonder whether the training parents receive on earth today is designed to be used in a future non-physical existence, when two once-human spirits become the protector and nurturer beings of lesser developed spiritual beings for an entire planet.

Returning to the illustration in Figure 49, which shows the interrelationship between the torus in a human and the torus of different celestial bodies, we now can see that this once-abstract idea is becoming more realistic given the double-torus structure's reality. The double-torus structure shows how the tori of multiple people, scattered across the planet's surface, can be integrated through the globe-shaped torus into the earth's donut-shaped torus. This concept also shows how human beings on earth could be linked to celestial objects in the macroscopic universe, as well as how the torus of subatomic particles can be linked through the intermediate steps of molecules, tissue and organs into the human body's torus. In a wave-based reality, literally everything in the universe—from subatomic to celestial scales—could be linked through the hierarchical torus structure!

APPENDIX D
Framework for
Spiritual Development

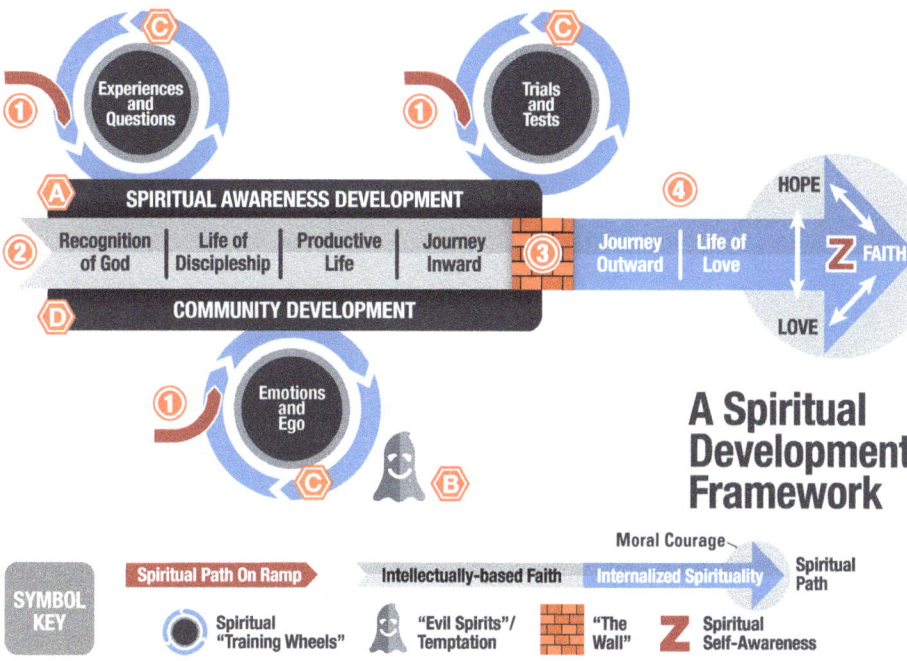

Figure 54: A spiritual development framework

It wasn't until after my spiritual experiences began that I realized how important it is to have a roadmap to guide me through the spiritual development process. After finding a framework that meshed well with what I had experienced, I embellished it and eventually produced the process laid out in Figure 54 and described in chapter nine. It was a work-in-progress for at least two years and has been updated many times over the course of writing my books. This certainly is not the only spiritual development framework available, although it has worked well for me and serves as a helpful reference to this day. Other frameworks, such as those I listed in chapter nine's further reading, can be found in the writings of theologians, monastics, mystics, and spiritual leaders across the ages. It's worth your time to find one that resonates with you.

The framework above is not as complicated as it might first appear. The diagram contains four broad steps, labeled numbers 1–4, that correspond to major stages of faith:

1 Entering the spiritual path

2 Developing an intellectual understanding of faith

3 Reconciling your intellectual understanding of faith with your life experiences

4 Internalizing and maturing a unique spirituality

The six stages of faith, shown in grey and blue within the arrow that stretches left to right in the center of Figure 54, represent different levels of spiritual maturity:

1. **Recognition of God** – acknowledging the possibility that there is something else to life and desiring a better understanding of it.

2. **Life of discipleship** – developing an intellectual understanding of a faith tradition by studying its scripture, learning its doctrine, understanding its history and traditions, and participating in its rituals and acts of service.

3. **Productive life** – applying what has been learned through discipleship studies and participating in acts of service while assuming spiritual leadership opportunities.

4. **Journey inward** – questioning some of what one intellectually believes because it conflicts with their life experiences. Significant life events, spiritual experiences, or philosophical questioning can spark movement into this stage. This stage culminates in *The Wall* where all these things are reconciled.

5. **Journey outward** – internalizing and rebuilding one's beliefs based on the unique spirituality that was just reconciled. What one believes here is stronger than it ever has been, and it is marked by an increase in one's joy and love for others.

6. **Life of love** – deepening a real love for others as one's spiritual development becomes more self-motivated and self-directed.

To learn more about these stages, read *The Critical Journey: Stages in the Life of Faith* by Janet Hagberg and Robert Guelich. *The Critical Journey* dedicates a chapter to each stage, describing its characteristics, how one moves through it, and how one can become "stuck" within it.

While it may not be apparent, life experiences are our spiritual development teachers. Their influence is illustrated in the development framework's four processes, which are labeled with letters A–D in Figure 54:

A Developing a spiritual practice

B Dealing with, and overcoming, temptation and fear

C Encountering (and learning from) opportunities for spiritual growth

D Developing a community that supports one's spirituality

The framework's four processes are discussed in my third book, if you would like additional details. Publication information can be found in the Bibliography.

It's important to note that the framework shown develops a "unique spirituality," because one's individual life experiences are foundational to the spiritual reconciliation process. The way each of us approaches spirituality—the path we take and the experiences we have—is as unique as our personalities, but the common thread is the personal transformation experienced as we grow to love others more deeply.[105] Let's examine this framework in greater detail to see how these steps and processes interact with each other. The four steps that follow were derived from the six stages of faith listed in the box above.

Step 1 – Entering the Spiritual Path

Most people cannot focus on spiritual things until their basic needs are met: food, shelter, and clothing. One person who recognized this was Father Bruce Ritter, the founder of Covenant House, a Manhattan-based Catholic charity. During a sermon in the 1980s, Father Ritter described how Covenant House

Figure 55: Step One - A spiritual path on-ramp

volunteers made sandwiches and hot chocolate to distribute each night to kids living on the streets of Manhattan. He said that people who lived on city streets had to be "brought up to a level" where they could be open to a spiritual message, but even after their physiological needs are met, people will not automatically launch into a spiritual journey.

Something must drive them to pursue it, which I call a "spiritual path on-ramp."

Entry points onto my framework's spiritual path are designated by a curved arrow, as shown in Figure 55. They represent significant life events where this realization could occur. Potential on-ramps could include an unexplained spiritual experience, a life-threatening illness, or the very lack of food, shelter and clothing. At this point, one's spiritual understanding will not have to be perfectly formed; however, a feeling should exist that there is more to life than what is seen, a condition that opens the door to spiritual exploration.

Step 2 – Developing an Intellectual Understanding of Faith

Figure 56: Step Two - Develop an intellectual understanding of faith

People typically learn about a faith tradition through participation in religious programs where they study spiritual basics (Figure 56). It's only after people can intellectually grasp the concept of a non-physical world that they'll begin to look for evidence of its existence. This understanding, however, must start with basic concepts, not complex ones. Equally as important, many religious institutions provide opportunities for its members to serve those in need, which help instill in them a desire to help others. As their intellectual understanding of faith deepens, they also may be interested in helping others who are new to the spiritual journey.

Step 3 – Reconciling Your Intellectual Understanding of Faith with Your Life Experiences

Figure 57: Step Three - Reconcile your faith with your life experiences

This step can be difficult, because the "intellectual" faith one once had been so certain of now is being questioned (Figure 57). Many people approach this step and back away because they are uncomfortable with the uncertainty of their struggle. Questioning aspects of one's faith is normal during this stage, however, and should cause no guilt or shame. It is a normal part of the spiritual development process. Even amidst uncertainty and doubt, people journeying through this step should not give up on their faith. They should instead understand that this process inevitably leads to the reconciliation and internalization of a unique spirituality they can believe wholeheartedly.

The reconciliation process is presented as a wall. Passing through *The Wall* is a milestone at the end of one's journey inward. It signifies that one's faith has been shaped and tested through their unique life experiences, and the individual emerges strong and confident.

On a personal note, I spent over a year between 2008 and 2009 trying to pass through my wall. I was able to move forward in 2009 only after dropping my belief in the Trinity, because at the time I could not reconcile or explain it. I was able to reincorporate this belief years later however, when I could explain it from a spiritual perspective. The truth is that you may have to give up a belief to move forward with your spiritual development, but you may yet reincorporate that belief as your spirituality matures … no matter how disconcerting it may feel to you early on.

Step 4 – Maturing a Unique Spirituality

Figure 58: Step Four - Mature a unique spirituality

People reaching this step (Figure 58), the last of the stages of faith, realize how their spiritual understanding has deepened. They have been transformed through internalization of a spirituality that now directs their lives. They also may find that, while they continue to rely on their faith's general teachings, specific doctrines and beliefs have less importance than before. They are now driven more by a focus on deepening relationships with those around them and on building strong, healthy communities.

Spiritual Development — An Individual Journey

"You can lead a horse to water, but you can't make it drink" is an old saying, and it's true for one's spiritual journey as well. Churches can offer programs, workshops, retreats, special emphases, and so forth, to promote one's spiritual development, but it is ultimately up to the individual to decide whether to pursue it. While churches can support people through all six stages of faith, their instructional work typically diminishes after stage three. The *Journey Inward* and certainly *The Wall* also are highly personal stages that may require more individualized spiritual direction. The latter two stages in step four are entirely personal. Not everyone will be interested in undertaking this work. But remember, the only person who can develop your spirituality is you.

Putting it all Together — Building Moral Courage

Developing an intellectually-based faith	THE WALL: Reconciling intellectual faith with life and spiritual experience	Realizing that the spirit world exists	Maturing one's love for others	Learning hope - how to persevere in faith and love	Integrating faith, hope and love to build moral courage

Juxtapositions of FAITH | Juxtapositions of LOVE FOR OTHERS | Juxtapositions of HOPE

Stand in the Mandorla

Figure 59: The role of juxtapositions in spiritual development

A *juxtaposition* is a side-by-side comparison of two contrasting objects—"junctures", if you will, intersections of concepts or ideas that help us make choices. We make choices throughout our lives that reflect what we believe in our minds or hold in our hearts at the time we make them. We use juxtapositions to help us organize, validate, or prioritize the many narratives surrounding those choices. Most of the time we make choices implicitly, like when deciding what foods to eat or what movie to see, but the comparison process becomes more obvious when we make choices that have long-term ramifications, such as deciding where to live or where to work. When making these comparisons, we rely on our research, education, and the relevant experiences of others to help us decide.

The Wall in the Spiritual Development Framework represents the first of many spiritual juxtapositions you will face in life. Spending time there to develop and internalize your faith is important because it will serve as the base from which your spirituality grows and changes during the rest of your life.

Figure 59, which expands on the items in Figure 54, shows three types of life juxtapositions: faith, love for others, and hope—and how they teach the principles of humility, selfless love, perseverance, and moral courage. We've already mentioned how internalized faith forms the baseline that later juxtapositions strengthen and mature. The second kind of juxtaposition, Love for Others, is encountered throughout our lives. These personal growth opportunities (choice juncture points) deepen our appreciation over time for the many opposites we encounter on earth. Distinctions in age, gender, religion, ideology, philosophy, and culture all underscore the many ways in which we can come to love each other in a selfless way … as who you are instead of what you are.

At the far end of the juxtaposition framework, after our experiences with faith and love, we tackle the juxtaposition of Hope, a struggle that teaches us how to maintain our faith and love for others in the midst of trial. It is through these juxtapositions that we develop perseverance and moral courage. The culmination of the process is the transformation of our worldview, represented by the golden *Stand in the Mandorla* icon on the right-hand side of Figure 59.[106] This is the point where our love for others surpasses our instinctual drive for self-preservation, which (my perspective) brings our earthly training to an end.

Why Develop Moral Courage?

As the three traits of faith, hope and love for others mature (Figure 60), we may find ourselves cultivating a sensitivity toward the suffering of others—despite any paradoxes that may remain within ourselves. This produces moral courage. In his book, *The Courage to Create*, Rollo May wrote that moral courage almost always has its beginning in the development of this type of sensitivity.[107] He also stated that the most prevalent form of cowardice today comes when we cut off our empathy toward the suffering of others, which leads to our unwillingness to get involved when we see them suffering.[108]

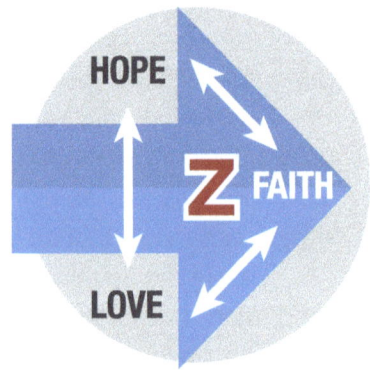

Figure 60: Moral courage - the integration of faith, hope and love for others

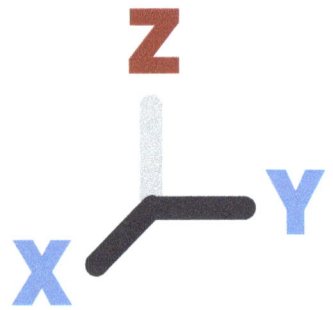

Figure 61: The "Z" axis

Culmination of the Process— Becoming Spiritually Self-Aware

In the last stage of spiritual development, we finally can envision meeting our ultimate objectives: to love others as yourself and to love our spiritual parent. This will require a worldview change—the realization that we are eternal spiritual beings. Reaching the highest level of spiritual maturity brings about "spiritual self-awareness," represented by the Z in the circle at the far end of the spiritual development framework arrow in Figure 60. The Z is a metaphor for climbing the z-axis, the vertical axis in the three-dimensional graph that represents rising above the temporal world (the x-axis and y-axis in Figure 61). If we can detach ourselves from earthly activities, we can observe and think about the effect of other people's emotions, egos, attitudes, and prejudices without being affected by them. While this technique could be practiced at an earlier stage of faith, I believe its impact cannot be fully realized until we reach the last stage.

After becoming spiritually self-aware, we are able to transfer the source of the love we receive from earthly people to spiritual beings—from our physical dimension friends and loved ones to our spiritual mentors. This change will enable you to selflessly love others here on earth even if you receive nothing in return—at best— or persecution and death—at worst. Why? Because the source of your love doesn't reside in the physical dimension—in fact, it no longer comes from mankind. The culmination of this process is your transformation into a spiritual warrior, someone who fights not with weapons that attack and kill, but with love and compassion that transform hearts. The mission of spiritual warriors is to bring love back into the hearts of people from whom it has disappeared, work that requires strength of heart, because it could take months or years—and many lives—to accomplish the objective.

You may ask, how could such a life test be given? If you are Jewish, or familiar with their history, you know that Jews have been persecuted consistently through the centuries. If you couch these events as the culmination of a spiritual development process, they could be seen as such a test. The same could be shown for the events in the Greek Bible book of Revelation, but I will not discuss them here. If you would like to read my perspective on Revelation as a spiritual test, it can be found in chapter thirteen of my third book. The publication information can be found in the Bibliography.

APPENDIX E
Wave-Based Explanations for Miracles in Bible Scripture

The intent of this appendix is to show that the majority of miracles recorded in Bible scripture can be explained as abilities currently known to parapsychologists. It supports my belief that everything that has happened, will happen and is happening now all obey the same laws of physics, which I contend are fully wave-based. It's not a miracle—it's [wave-based] science!

Telepathy: Mentally Communicating Images or Thoughts Between Two or More Beings

Universal Hearing – Where a group of people hear the same message but in their own language
- Acts 2:5–12 – On Pentecost, Peter addresses a crowd of Jews from many nations in their native language which could have been a demonstration of telepathic communication.

Spirit Communications – Where a non-physical being communicates with a human
- Genesis 22:11–12 – An angel from heaven speaks to Abraham.
- Exodus 3:4 – God calls Moses from a burning bush.
- Joshua 5:13–14 – Joshua speaks with the angel who commands God's army.
- Judges 6:20 – An angel speaks to Gideon.
- Judges 13:3 – An angel tells Manoah his wife will give birth to Samson.
- 1 Samuel 2:27 – God reveals himself to Eli's family.
- 1 Samuel 3:10 – God speaks to Samuel.
- 1 Samuel 9:15–20 – God tells Samuel to anoint Saul king of Israel.
- 1 Kings 19:7, 13 – God speaks to Elijah.
- Daniel 4:31 – God speaks to King Nebuchadnezzar.

- Daniel 8:18 – God speaks to Daniel.
- Ezekiel 4 – God tells Ezekiel about the siege of Jerusalem.
- Jeremiah 1 – God calls Jeremiah to speak for him to the people of Israel.
- Matthew 17:3 – Moses and Elijah speak with Jesus at his transfiguration.
- Luke 1:28 – An angel speaks to Mary.
- Luke 2:9 -14 – An angel speaks to shepherds.
- Luke 9:35 – God speaks at Jesus's baptism.
- John 12:28 – God speaks from heaven.
- Acts 8:26 – An angel speaks to Philip.
- Acts 9:4–7 and Acts 26:12–18 – Jesus speaks to Saul (Paul) on the road to Damascus.
- 2 Peter 1:21 – Prophecy had its origin in the Holy Spirit speaking through prophets.

Spirit Communication via Trance – Humans receiving a communication through a trance or vision

- Numbers 24:4 (King James Version) – Balaam's third oracle, blessing the Israelites.
- Jeremiah 1:11–14 – God trains the prophet Jeremiah in how to interpret visions. Interestingly, Charles Leadbeater discusses similar training a master gives their students on how to interpret things they see in the astral dimension.[109]
- Acts 10:9–16 – God speaks to Peter while he is in a trance, saying all meat is edible.
- Acts 22:17–21 – God sends Paul to teach the Gentiles.

Spirit Communication via Dreams – Humans receiving a communication via a dream

- Numbers 12:6 – "He [God] said [to Aaron and Miriam], 'Listen to my words: When a prophet of the Lord is among you, I reveal myself to him in visions, I speak to him in dreams.'"
- Genesis 28:12 – Jacob sees angels ascending and descending a stairway that reaches to heaven.

- Genesis 31:24 – God communicates with Laban in a dream.
- Genesis 37:5 – Joseph has dreams about his ruling over his eleven brothers, the sons of Jacob.
- Genesis 41:17 – Pharaoh has dreams of years of plenty then famine, which Joseph interprets.
- Daniel 10:7-9 – Daniel sees a vision.
- Job 33:14–15 – God speaks to people in dreams and visions.

Remote Perception: Projecting One's Conscious Energy to Perceive Remote Surroundings

- 2 Kings 6:8–12 – During a war with the king of Aram, the prophet Elisha tells the king of Israel about the king of Aram's plans, enabling him to avoid the king of Aram's traps.
- John 1:48 – Jesus remotely observes Nathaniel under a fig tree.

Energy Healing: Healing a Person's Etheric Body, Which in Turn Heals Their Physical Body

Jesus's Direct and Remote Healing

- Matthew 8:2–4 – Man with leprosy.
- Matthew 8:5–13 – Remote healing of a Roman centurion's servant.
- Matthew 8:14–15 – Peter's mother-in-law's fever.
- Matthew 9:2–7 – Paralytic on a mat.
- Matthew 9:20–22 – Woman with bleeding for many years, who touched Jesus's cloak.
- Matthew 9:27–31 – Two blind men.
- Matthew 12:10–13 – Man with shriveled hand.
- Matthew 15:21–28 – Remote healing of Canaanite woman's daughter.
- Matthew 20:29–34 – Two blind men.
- Mark 2:3–12 – Paralytic who was lowered from a hole in the roof of a house.
- Mark 7:31–37 – Deaf mute.
- Mark 10:46–52 – Blind beggar named Bartimaeus.

- Luke 13:11–13 – Crippled woman.
- Luke 14:1–4 – Man with dropsy.
- Luke 17:11–19 – Remotely healed ten men with leprosy.
- Luke 22:50–51 – Restores the ear of the high priest's servant.
- John 4:46–54 – Remotely heals an official's son at Capernaum.
- John 5:1–9 – Invalid at the Pool of Bethesda.

Paul's Healing
- Acts 19:11–12 – Handkerchiefs that touched Paul could cure the sick.

The Apostles' Healing
- Acts 5:15–16 – The apostles performed many healings while in Jerusalem.

Psychokinesis: The Manipulation or Manifestation of Physical Objects

Jesus's Command over Forces of Nature and the Manipulation of Material Objects, Such as Food
- Matthew 8:23–27 – Calms a storm while in a boat on a lake.
- Matthew 14:25 – Walks on water.
- Matthew 14:15–21 – Feeds 5,000+ people from two fish and five loaves.
- Matthew 15:32–38 – Feeds 4,000+ people from seven loaves and a few fish.
- Matthew 17:24–27 – Predicts finding a coin in a fish's mouth to pay the synagogue tax.
- Matthew 21:18–22 – Withers fig tree.
- John 2:1–11 – Turns water into wine.

Bodily Protection from Fire
- Daniel 3:25, 27 – An angel protects Shadrach, Meshach, and Abednego in the fire. Charles Leadbeater writes in *The Astral Plane* how someone can handle fire without harm. He said

that the application of a thin layer of the etheric dimension's elemental essence around one's hands will enable them to pick up burning coal or a red-hot iron in safety.[110]

Body Levitation
- Judges 13:20 – An angel ascends to heaven in the smoke from a fire.
- Ezekiel 8:3 – A spirit transports Ezekiel to Jerusalem.
- Matthew 14:29 – Peter walks on the water toward Jesus.
- Acts 1:9 – Jesus is taken up into heaven.

Spiritual Manifestations Involving Physical Objects
- Genesis 15:17 – God makes a covenant with Abram (Abraham) by passing a smoking firepot and a blazing torch between pieces of meat.
- Exodus 3:2 – Moses sees a burning bush that was not being consumed.
- Exodus 13:21 – A pillar of cloud (by day) and of fire (by night) guides the Israelites in the wilderness.
- Exodus 40:34 – A cloud covers the tabernacle in the wilderness, a sign of God's presence.
- Leviticus 9:24 – A pillar of fire consumes the first burnt offering offered by the Israelite priests.
- Numbers 16:31-34 – The ground opens to consume Korah and his family after his disobedience.
- Numbers 16:35 – Fire consumes 250 men who inappropriately burned incense to God.
- Judges 6:21 – An angel touches meat with his staff and fire flares from a rock to consume it.
- 1 Kings 18:38 – At Elijah's request, fire from heaven consumes an offering to God.
- 2 Kings 1:10, 12 – At Elijah's request, fire from heaven consumes two separate groups of fifty soldiers.
- 1 Chronicles 21:26 – At King David's request, fire from heaven consumes a burnt offering.

- Acts 2:3 – Tongues of fire appear over the heads of Jesus's twelve apostles on the day of Pentecost.
- Acts 12:7 – An angel frees Peter from Herod's prison in Jerusalem.

Non-local Consciousness: Accessing Spiritual Abilities or Traveling Outside the Physical Body

Out-of-Body (OBE) or Near-Death (NDE) Experiences

- Isaiah 6:1–8 – Isaiah describes a heavenly temple and replies, "Here am I, send me!" when God asks for a volunteer while he is present. (OBE)
- Ecclesiastes 12:6 – A reference to the "silver cord being broken." Jurgen Ziewe and other OBE explorers mention a "silver cord" that was attached to their etheric body during their OBE travels.[111] This passage refers to the silver cord being cut when your physical body dies.
- 2 Corinthians 12:2–4 – Paul describes what could have been a near-death experience that occurred after one of his life-threatening confrontations, such as when he was stoned and left for dead at Lystra in Acts 14:19–20. (NDE)
- Revelation 1:10, 4:2, 17:3 and 21:10 – John travels "in the spirit" during an out-of-body experience to observe the events he described while in heaven. (OBE)

Humans Accessing Higher Dimensions

- Exodus 34:30 – Moses's face is so radiant that he must wear a veil when among the Israelites. (Upper Mental dimension)
- Matthew 17:1–7 – Jesus's transfiguration. (Upper Mental dimension – my opinion only)
- John 4:16, 18 – Jesus knows the woman at the well's marital history. (Metaphysical texts say that a person's life history is stored in the Akashic records in the Mental dimension.)

Instances of Dead Humans Being Brought Back to Life

- 1 Kings 17:17–24 – Elijah revives the son of the widow at Zarephath.
- 2 Kings 4:8–37 – Elisha lies on top of the Shunammite woman's son to revive him.
- Matthew 9:18–19, 23–26; Luke 8:41–42, 49–56 – Jairus's daughter. Jesus tells Jairus in Luke 8:50 to "believe, and she will be healed," which is indicative of the role of intention in changing the outcome of life events.
- Luke 7:11–15 – Jesus revives a widow's son at Nain.
- John 11:1–44 – Jesus revives Lazarus who had been dead for four days.
- Acts 9:36–42 – Peter revives Tabitha, also called Dorcas, in Joppa.
- Acts 20:7–12 – Paul revives Eutychus, who fell from a third story window in Troas.

Non-Physical Beings Appearing to Humans

- Genesis 3:8 – God appears to Adam and Eve as he walks through the Garden of Eden.
- Genesis 18:1, 2 – God appears to Abraham near the great trees of Mamre.
- Genesis 32:24 – Jacob wrestles with an angel.
- Exodus 24:9, 10 – Moses, Aaron, and the elders of Israel see God.
- 1 Samuel 28:13, 14 – Saul has a medium call up the spirit of Samuel.
- Daniel 5:5 – A disembodied human hand appears and writes on a wall, foretelling a king's death. This phenomenon can occur during a physical mediumship session.
- Daniel 9:21 – Daniel sees the Angel Gabriel.
- John 20:12, 14, 19 – Jesus and two angels appear to the disciples.
- Luke 24:15, 30, 31, 36 – Jesus appears on the road to Emmaus.
- John 21:1 – Jesus appears to His disciples at the Sea of Galilee.
- Acts 10:3, 19 – The Roman centurion Cornelius sees an angel.

BIBLIOGRAPHY

Adams, Keli. *Drop the BS (Belief Systems) and Be.* Bloomington, IN: AuthorHouse, 2019.

Babbitt, Edwin D. *The Principles of Light and Color.* 1878 ed. London: Forgotten Books, 2018.

Backster, Cleve. *Primary Perception: Biocommunication with Plants, Living Foods, and Human Cells.* White Rose Millennium Press, 2003.

Barker, Kenneth L. et. al. *The NIV Study Bible: New International Version.* Grand Rapids, MI: Zondervan Bible Publishers, 1985.

Bengston, William, and Sylvia Fraser. *The Energy Cure: Unraveling the Mystery of Hands-on Healing.* Boulder, CO: Sounds True, 2011.

^Bentov, Itzhak, and Mirtala Bentov. *From Atom to Cosmos.* DVD. Venice, CA: New Science Ideas, 2005.

Bentov, Itzhak. *Stalking the Wild Pendulum: On the Mechanics of Consciousness.* Rochester, VT: Destiny Books, 1988.

Bentov, Itzhak. *A Brief Tour of Higher Consciousness: A Cosmic Book on the Mechanics of Creation.* Rochester, VT: Destiny Books, 2000.

*Besant, Annie, C. W. Leadbeater, Elizabeth Preston, and C. Jinarājadāsa. *Occult Chemistry: Investigations by Clairvoyant Magnification into the Structure of the Atoms of the Periodic Table and of Some Compounds* (Third Edition). Adyar, Madras, India: Theosophical Pub. House, 1951. (Available in PDF form on the Internet).

Besant, Annie, C. W. Leadbeater. *Thoughtforms*, Wheaton, IL: Quest Books, 1901.

^Bladon, Lee. *The Science of Spirituality*. Self-published. www.evolvingsouls.com, 2007.

Brooks, Mark Hunter. *Christianity from a Different Perspective: Real Spirituality in a Quantum World (Version 1.0)*. Charlotte, NC: From a Different Perspective LLC, 2016.

Brooks, Mark Hunter. *Christianity from a Different Perspective: Real Spirituality in a Quantum World – Version 2.0*. Charlotte, NC: From a Different Perspective LLC, 2017.

Brooks, Mark Hunter. *Christianity from a Spiritual Perspective: Real Spirituality in a Quantum World – Version 3.0*. Charlotte, NC: From a Different Perspective LLC, 2019.

Brooks, Mark Hunter. "A Multidimensional Explanation for Magnetism." *Quest* 108:3, pg 34-38. Wheaton, IL: Theosophical Society in America, 2020.

Caplan, Seth, prod. *Flatland: The Movie*. DVD. Austin, TX: Flat World Productions, 2007. http://flatlandthemovie.com/.

Cardeña, Etzel, John Palmer, and David Marcusson-Clavertz, eds. *Parapsychology: A Handbook for the 21st Century*. Jefferson, NC: McFarland & Company, Publishers, 2015.

Coleman, Tim, director. *The Afterlife Investigations: The Scole Experiments*. UFOTV. 2011, https://www.imdb.com/title/tt11156348/.

Cramer, John. *The Quantum Handshake: Entanglement, Nonlocality and Transactions*. Switzerland: Springer International Publishing, 2016.

Dawkins, Richard. *The Blind Watchmaker*. New York: Norton, 1986.

Eeman, L.E. *Co-operative Healing: The Curative Properties of Human Radiations.* London: Author Partner Press, 1947.

Emoto, Masaru. *The Hidden Messages in Water.* New York: Atria Books, 2005.

Gherwal, Singh Rishi. *Great Masters of the Himalayas: Their Lives and Temple Teaching.* Volume One, Third Edition. Self-Published, 1927.

Glashow, Sheldon Lee, and Ben Bova. *Interactions: A Journey through the Mind of a Particle Physicist and the Matter of This World.* New York, NY: Warner Books, 1989.

Grof, Stanislav, and Christina Grof (editors). *Spiritual Emergency: When Personal Transformation Becomes a Crisis.* New York: Jeremy P. Tarcher / Putnam, 1989.

Hagberg, Janet, and Robert A. Guelich. *The Critical Journey: Stages in the Life of Faith.* Salem, WI: Sheffield Pub., 2005.

Hardy, Alister. "Biology and Psychical Research." In the *Proceedings of the Society for Psychical Research.* 50. Society for Psychical Research, May 1953.

Hawkins, Greg L., and Cally Parkinson. Move: *What 1,000 Churches Reveal about Spiritual Growth.* Grand Rapids, MI: Zondervan, 2011.

Heriot, Drew, Director; Rhonda Byrne, Author. *The Secret.* Prime Time Productions, 2006.

*Keen, Montague, Arthur Ellison, and David Fontana. *The Scole Report.* England: Saturday Night Press Publications. 2011. http://www.snppbooks.com. Previously published as *Proceedings of the Society for Psychical Research.* Volume 58, part 220. November 1999.

Kilner, Walter J. *The Human Atmosphere: Or the Aura Made Visible by the Aid of Chemical Screens* (Classic Reprint). 1911 ed. London: Forgotten Books, 2015.

^Laurency, Henry T. *The Knowledge of Reality.* Skövde, Sweden: Henry T. Laurency Publishing Foundation, 1979.

^Laurency, Henry T. *The Philosopher's Stone.* Skövde, Sweden: Henry T. Laurency *Publishing Foundation,* 1985.

*Leadbeater, Charles W. *The Astral Plane: Its Scenery, Inhabitants and Phenomena* (Classic Reprint). Forgotten Books, 2015.

Leadbeater, Charles W. *The Devachanic Plane: Its Characteristics and Inhabitants* (Classic Reprint). Forgotten Books, 2015.

Leadbeater, Charles W. *Invisible Helpers* (C.W. Leadbeater Collection: Volume 2). CreateSpace Independent Publishing Platform, 2016.

Leadbeater, Charles W. *Man Visible and Invisible.* Theosophical Pub. House, 1971.

Long, Max Freedom. *The Secret Science at Work.* Santa Monica, CA: DeVorss & Company, 1953.

Maraldi, Everton de Oliveira. *Parapsychology and Religion.* Leiden, The Netherlands: Brill, 2021.

May, Edwin C. et. al. *ESP Wars East & West: An Account of the Military Use of Psychic Espionage as Narrated by the Key Russian and American Players.* Hertford, NC: Crossroad Press, 2015.

May, Rollo. *The Courage to Create.* New York: Norton, 1975.

McGinn, Bernard. *The Essential Writings of Christian Mysticism.* New York: Modern Library, 2006.

McTaggart, Lynne. *The Power of Eight: The Miraculous Healing Power of Small Groups.* Atria Books, 2017.

Mead, Carver A. *Collective Electrodynamics: Quantum Foundations of Electromagnetism.* Boston: MIT Books, 2000.

^Meek, George W. *After We Die, What Then?* Columbus, Ohio: Ariel Press, 1987.

Mishlove, Jeffrey. *The PK Man.* Charlottesville, VA: Hampton Roads Publishing Company, 2000.

Monroe, Robert A. *Ultimate Journey.* New York: Doubleday, 1994.

Morgan, Marlo. *Mutant Message Down Under.* New York NY: HarperCollins, 1994.

Patten, Leslie with Terry Patten. *Biocircuits: Amazing New Tools for Energy Health.* Tiburon, CA: HJ Kramer Inc., 1988.

Phillips, Stephen M. *Extrasensory Perception of Quarks.* Wheaton, Ill: Theosophical Pub. House, 1980.

Phillips, Stephen M. *ESP of Quarks and Superstrings.* New Delhi: New Age International, 1999.

Powell, Arthur E. *The Astral Body and other Astral Phenomena.* Wheaton, Illinois: The Theosophical Publishing House, 1927.

Powell, Arthur E. *The Causal Body and the Ego, Second Edition.* Adyar: The Theosophical Publishing House, 2016.

*Powell, Arthur E. *The Etheric Double: The Health Aura.* Wheaton, Illinois: The Theosophical Publishing House, 1925.

Powell, Arthur E. *The Mental Body.* London: The Theosophical Publishing House Limited, 1927.

^Powell, Arthur E. *The Solar System.* Wheaton, Illinois: The Theosophical Publishing House, 1930.

Primack, J. R., and Nancy Ellen Abrams. *The View from the Center of the Universe: Discovering Our Extraordinary Place in the Cosmos.* New York: Riverhead Books, 2007.

Rhine, J.B. "The Parapsychology of Religion: A New Branch of Inquiry," in Coly, *Parapsychology, Philosophy and Religious Concepts*, 192-215.

Russell, Walter. *The Universal One.* Waynesboro, Virginia: The University of Science and Philosophy, 1928.

Sadhu, Mouni. *Concentration: A Guide to Mental Mastery.* Hollywood, CA: Melvin Powers Wilshire Book Company, 1959.

Schlitz, Marilyn, Cassandra Vieten, and Tina Amorok. *Living Deeply: The Art and Science of Transformation in Everyday Life.* Petaluma, CA: Noetic Books, 2009.

Spiritual Science Founders Association. Antequera, Andalucía, Spain. http://ssf-robin-foy.com.

Sinclair, Upton. *Mental Radio.* Charlottesville, VA: Hampton Roads, 2001.

Stodolna, A.S. et. al. "Hydrogen Atoms under Magnification: Direct Observation of the Nodal Structure of Stark States." *Physical Review Letters,* 110, 213001 – published 20 May 2013. DOI: https://doi.org/10.1103/PhysRevLett.110.213001

White, John and Stanley Krippner, eds *Future Science: Life Energies and the Physics of Paranormal Phenomena.* Garden City, NY: Anchor Press/Doubleday, 1977.

Wolchover, Natalie. "Neutron Lifetime Puzzle Deepens, but No Dark Matter Seen." *Quanta Magazine.* 13 February 2018.

Wood, Ernest. *Concentration: An Approach to Meditation.* Wheaton, IL: Theosophical Publishing House, 1949.

*Wolff, Milo. *Schrödinger's Universe: Einstein, Waves & the Origin of the Natural Laws.* Parker, CO: Outskirts Press, 2008.

Ziewe, Jurgen. *Multidimensional Man.* Raleigh, NC: Lulu Press, 2008.

* Highly recommended reading
^ Works that discuss our eternal existence

INDEX

The Index cites words appearing in the text or in substantive endnotes. Words appearing in the Table of Contents, Table of Figures, Bibliography, book titles, illustrations or captions under figures are not cited.

A

Abrams, Nancy Ellen 67
Adams, Keli 60
Anu 15, 22, 30, 31, 46, 47, 48, 49, 50, 51, 97, 99, 100, 102, 106, 108, 110, 125
Astral 15, 16, 23, 41, 44, 45, 47, 48, 49, 50, 60, 61, 62, 63, 64, 70, 80, 85, 94, 97, 105, 106, 110, 111, 125, 127, 151, 153
Aura 23, 44, 47, 48, 102, 103, 106, 108

B

Babbit, Edwin 30
Backster, Clive 171
Barker, Kenneth 157
Bengston, Bill 94, 95
Bentov, Itzhak (Ben) 171
Bentov, Mirtala 171
Besant, Annie 21, 22, 30, 37, 124, 126, 172
Bladon, Lee 108

C

Caplan, Seth 172
Cardena, Etzel 170
Causal 42, 99, 102, 114
Chakra 15, 23, 26
Concentration 34, 89, 111, 139
Cramer, John 119

D

Dawkins, Richard 83, 84

E

Eeman, Ernst 95
Emoto, Masaru 86, 87
Etheric 41, 44, 45, 47, 49, 50, 55, 57, 59, 60, 63, 64, 85, 90, 92, 93, 95, 97, 98, 106, 107, 108, 110, 111, 123, 124, 125, 126, 152, 154, 155, 168

F

Falsify 19, 118, 119, 121, 125, 126, 128
Framework 15, 18, 19, 34, 71, 74, 75, 76, 78, 118, 141, 142, 143, 144, 147, 148

G

Gherwal, Singh Rishi 168
Glashow, Sheldon 67
Grof, Stanislav 14, 82
Guelich, Robert 142

H

Hagberg, Janet 142
Hardy, Alister 90
Hawkins, Greg 171
Healing 18, 54, 55, 56, 62, 71,
    81, 86, 87, 90, 91, 92, 93,
    94, 95, 111, 123, 152, 153
Heriot, Drew 159

I

Infinity 65, 66, 67, 100, 111
Interpenetration 42, 44, 51,
    63, 109

K

Keen, Montague 173
Kilner, Walter 47

L

Laurency, Henry 108
Leadbeater, Charles 21, 22, 30,
    37, 41, 47, 59, 64, 124, 126,
    127, 151, 153
Long, Max Freedom 113

M

Maraldi, Everton 54
Mass 17, 38, 42, 58, 59, 65,
    105, 106, 108, 109, 110, 126

May, Edwin 55
May, Rollo 148
McGinn, Bernard 74
McTaggart, Lynne 95
Mead, Carver 119, 120
Meek, George 108
Mental 33, 41, 42, 44, 56, 64,
    73, 82, 85, 91, 104, 106,
    112, 127, 128, 155
Metaphysical 15, 16, 20, 24,
    41, 42, 44, 47, 84, 88, 111,
    116, 117
Milton and Wiseman 170
Mishlove, Jeffrey 56
Monroe Institute 169, 170
Monroe, Robert 127
Morgan, Marlo 55

N

Near-Death Experience 56, 80,
    155
Neutron 22, 50, 98, 109, 110,
    121, 122, 130, 134
Non-physical 14, 16, 17, 18, 19,
    21, 25, 26, 37, 40, 54, 58,
    60, 61, 62, 63, 71, 72, 73,
    79, 80, 85, 89, 93, 94, 98,
    100, 102, 105, 108, 110, 112,
    114, 115, 116, 127, 139, 150,
    156

O

Out-of-Body Experience 41, 59,
    60, 127, 155
Ouroboros 67

P

Patten, Leslie 95
Phillips, Stephen 167
Physical 13, 15, 17, 19, 33, 37, 40, 41, 42, 44, 47, 49, 51, 52, 53, 55, 56, 57, 60, 63, 64, 69, 71, 73, 80, 83, 88, 89, 90, 91, 92, 95, 98, 99, 100, 105, 106, 108, 110, 111, 120, 123, 125, 126, 127, 128, 129, 136, 137, 149, 152, 153, 154, 155, 156, 168
Powell, Arthur 21, 22, 37, 41, 64, 95, 173
Primack, Joel 67

Q

Quantum Disentanglement 17, 44, 45, 99, 100, 102, 110
Quantum Entanglement 16, 24, 44, 45, 56, 90, 93, 94, 167

R

Rhine, J.B. 53, 54
Russell, Walter 29, 30

S

Sadhu, Mouni 73
Scale 15, 17, 44, 46, 66, 67, 68, 69, 100, 111, 115, 120, 125, 136, 140, 168
Schlitz, Marilyn 174
Schrödinger, Erwin 29
Secret, The 85, 89
Spiritual Science Founders Association 129
Stodolna, A.S. 174

T

Telepathy 18, 54, 55, 59, 60, 112, 150, 171
Theosophical 15, 21, 22, 30, 37, 39, 41, 42, 45, 47, 48, 64, 83, 85, 100, 104, 112, 114, 116, 127, 172
Thoughtform 84, 85, 86, 87, 89, 90, 92, 93, 95, 105, 106, 108, 172
Torus 15, 19, 22, 23, 29, 30, 68, 69, 118, 125, 136, 137, 138, 139, 140, 167

W

Wave-based 5, 13, 17, 19, 20, 24, 25, 28, 29, 30, 31, 37, 40, 41, 42, 43, 44, 45, 49, 51, 54, 55, 56, 57, 58, 59, 63, 64, 90, 93, 96, 97, 98, 99, 104, 109, 110, 114, 118, 119, 120, 121, 122, 125, 126, 127, 130, 140, 150
Wolchover, Natalie 173
Wolff, Milo 29, 30, 49, 98, 119, 120, 121, 122, 123, 167
Wood, Ernest 73

Z

Ziewe, Jurgen 155

ENDNOTES

1   Wolff, Schrödinger's Universe, v.

2   Schlitz, Living Deeply, 206–207.

3   Besant and Leadbeater, Occult Chemistry, 13.

4   Leadbeater, The Astral Plane, 7–8.

5   Phillips, The ESP of Quarks, 237.

6   The initial experiments by Alain Aspect in this area in 1981 tested Bell's Inequality, which sought to prove that quantum entanglement was real and not a result of "hidden variables" from conventional physics. The last major issue was addressed in 2015 by Ronald Hanson of Delft University of Technology. Sourced from http://www.nature.com/news/quantum-spookiness-passes-toughest-test-yet-1.18255 on 16 December 2016.

7   Wolff, Schrödinger's Universe, v.

8   https://www.goodreads.com/author/quotes/278. Nikola_Tesla. Accessed on 6 June 2021.

9   Russell. The Universal One, 99.

10   Wolff, Schrödinger's Universe, 2, 21, 41. Wolff never considered that these drawings could be of a torus.

11   The 1908 book was the first of three editions. The other two were published in 1919 and 1951.

12   Occult is one of those terms that has acquired considerable extra baggage in modern times, thanks in large part to popular entertainment's sensational use of the term. It simply means "hidden" as in "hidden science" or "hidden spirituality."

13   See the YouTube video titled Visualizing 4D Geometry – A Journey into the 4th Dimension [Part 2], starting at 12:58 into the video. Sourced from https://www.youtube.com/watch?v=4URVJ3D8e8k&t=971s on 6 November 2018.

14   See comments at https://aciste.org/about-stes/common-challenges-following-an-ste/ , sourced on 21 November 2021.

15   These books are listed in the Bibliography. They are available on www.amazon.com. Search by my name—Mark Hunter Brooks.

16   See the YouTube video describing the demonstration at https://www.youtube.com/watch?v=gKCN1o6aJuY, sourced on 15 December 2016.

17   Lincoln, Don. The Origins of Mass, YouTube.com. Sourced from https://www.youtube.com/watch?v=x8grN3zP8cgon on 15 June 2017.

18   Watch the video Have you Ever Seen an Atom? to see how far technology has advanced in our ability to observe objects at the atomic scale. Sourced from https://www.youtube.com/watch?v=yqLIglaz1L0 on 12 November 2018.

19   Leadbeater, The Astral Plane, 125.

20   Gherwal, Great Masters of the Himalayas, 112–114.

21   Powell, The Etheric Double, 3. Leadbeater, The Astral Plane, 127—mentions the degree to which the physical and etheric bodies are entangled.

22   Leadbeater, The Astral Plane, 18.

23   Leadbeater, The Astral Plane, 16.

24   Leadbeater, The Astral Plane, 26.

25   Leadbeater, The Devachanic Plane. 76–79.

26   Powell, The Causal Body and the Ego.

27   Leadbeater, The Devachanic Plane. 79–80.

28   Ron Cowen (23 December 2015) The Quantum Source of Space-Time, Nature. 527(7578). Sourced from http://www.nature.com/news/the-quantum-source-of-space-time-1.18797 on 12 August 2016.

29   George Musser (19 January 2016) Quantum Weirdness Now a Matter of Time, Quanta Magazine. Sourced from https://www.quantamagazine.org/20160119-time-entanglement/ on 12 August 2016.

30   Leadbeater, The Astral Plane, 10.

31   Besant and Leadbeater, Occult Chemistry, 13.

32   Besant and Leadbeater, Occult Chemistry, 14.

33   Besant and Leadbeater, Occult Chemistry, 10.

34   Besant and Leadbeater, Occult Chemistry, 13.

35   Brooks. Quest.

36   The magnetic field lines illustration was sourced from the Internet at the link http://www.sciencekids.co.nz/pictures/physics/ironfilings.html on 21 November 2019. This site contains a large repository of free resources to assist educators teaching math, science and English.

37   See the link http://hyperphysics.phy-astr.gsu.edu/hbase/Waves/string.html, sourced on 15 June 2021.

38   Rhine, J.B. PhD. "The Parapsychology of Religion: A New Branch of Inquiry," in Coly, Parapsychology, Philosophy and Religious Concepts, 192–215.

39   Rhine, J.B. PhD. "The Parapsychology of Religion: A New Branch of Inquiry," in Coly, Parapsychology, Philosophy and Religious Concepts, 192–215.

40   Maraldi, Everton. Parapsychology and Religion.

41   These statistics were presented in Dean Radin's Monroe Institute Professional Seminar in 2016—https://www.youtube.com/watch?v=CEVah39GPuw (starting at 15:00), sourced on 7 December 2021. Per Radin in his presentation, the data appeared in the Psychological Bulletin in 1994, 1999, 2001, and 2010. The most recent meta-analysis data is

reviewed in the article "Explicit Anomalous Cognition: A Review of the Best Evidence in Ganzfeld, Forced-Choice, Remote Viewing and Dream Studies," in Cardena, et. al, Parapsychology: A Handbook for the 21st Century, pages 192–214. Additionally, this article responds to a meta-analysis performed by Milton and Wiseman in 1999, which has often been cited by skeptics as showing that psi does not exist. It points out that their study had a low number of participants who exhibited psi-conduciveness. When a subset of their data that met this criterion was isolated (about 10% of the data), the results showed a 34.2% hit rate.

42 Morgan, Mutant Message Down Under, pages 61–64 (chapter 8).

43 A photon can be described as a tiny burst, or quantum, of light.

44 Go to https://www.youtube.com/watch?v= d2TDXKfBaMQ to watch a video showing how this is done.

45 Leadbeater, Invisible Helpers.

46 Soul retrieval is taught in the Monroe Institute's Lifeline course. Sourced at https://www. monroeinstitute.org/products/lifeline-virtual-retreat on 1 January 2022.

47 Adams, Drop the BS (Belief Systems) and Be, pages 31–58 (chapters 6–9).

48 To get a better idea of what this looks like in three dimensions, you may want to watch this animation of a moving tesseract, sourced on 5 July 2021 at https:// en.wikipedia.org/wiki/File:Tesseract.gif (public domain animation created by Jason Hise).

49 See Scale of the Universe2 at http://htwins.net/ scale2/ , sourced on 17 April 2017.

50 Glashow, Interactions, 311.

51 Dark matter is a hypothetical form of matter thought to account for about 27% of the matter in the universe. In contrast, the matter in all the stars and galaxies only makes up about 5% of the matter in the universe.

52 Primack, View from the Center of the Universe, 160. Ben Bentov's other book, A Brief Tour of Consciousness, and his wife Mirtala's video, From Atom to Cosmos, are also good.

53 Bentov, Stalking the Wild Pendulum, 136–137.

54 See the trailer at https://www.youtube.com/watch?v=C8oiwnNlyE4 , sourced on 1 November 2021.

55 I first saw this saying as the title of a book written by David P. Campbell.

56 Hagberg, The Critical Journey.

57 Hawkins, Move.

58 McGinn, The Essential Writings of Christian Mysticism, Section 5.

59 Underhill, Mysticism, Part Two.

60 Powell, The Etheric Double, 63–66.

61 Powell, The Etheric Double, 64.

62 Powell, The Etheric Double, 63.

63 Leadbeater, The Astral Plane, 66.

64 Dawkins, The Blind Watchmaker, 46–48.

65 Powell, The Solar System, pages 50–54.

66 Backster, Primary Perception.

67 We earlier discussed this concept with telepathy. One of the first books published on it was popular American author Upton Sinclair's 1930 book Mental Radio, in which he evaluated his wife's (Mary Craig Sinclair) telepathy and clairvoyance abilities.

68 Emoto, The Hidden Messages of Water, 64–65.

69 Leadbeater, The Astral Plane, 8–9.

70 You can view the film's trailer at https://www.youtube.com/ watch?v=san61qTwWsU, sourced on 3 November 2021.

71 The best book I found on thoughtforms was the Theosophical Society book of the same title, written by Annie Besant and Charles W. Leadbeater. It is listed in the Bibliography.

72 Hardy, "Biology and Psychical Research," 114.

73 Long, The Secret Science at Work. Pages 14–48 (Chapters II and III).

74 This was a premise behind Seth Caplan's animated film Flatland. Also, see Leadbeater, The Astral Plane, 10 for an astral sight comment.

75 See Robert Miller (January 1976) The Energies of Spiritual Healing, Science of Mind, pages 22-27. This work was sponsored by the Ernest Holmes Research Foundation. These experiments are also mentioned in Chapter 33 (pages 431-444) of White, Future Science.

76 Besant and Leadbeater, Occult Chemistry, 13.

77 Leadbeater, The Astral Plane, 9.

78 Besant and Leadbeater, Occult Chemistry, 14.

79 Leadbeater, The Astral Plane, 29-106, and Leadbeater, The Devachanic Plane, 30-100.

80 This object's shape resembles the shape of a pomegranate, with its tiny opening at the top of the fruit.

81 Powell, The Causal Body and the Ego, 99.

82 1 Kings 7:20 and Exodus 28:33

83 Powell, The Causal Body and the Ego, 66–69.

84 Wolff, Schrödinger's Universe, 19.

85 See how others characterized atoms in chapter one's further reading.

86 American Spectator, September / October 2001, Volume 34 Issue 7, page 68. Sourced at http://worrydream.com/refs/ Mead%20-%20American%20Spectator%20Interview.html on 3 January 2022.

87 Wolff, Schrödinger's Universe, 130-132.

88 Wolchover, Quanta Magazine.

89 Wolff, Schrödinger's Universe, 108.

90 Annals of the NY Academy of Science. 480:304.

91 Arthur Powell's book The Etheric Double contains a number of drawings showing how pranic energy comes into the human body and how it is moved within it. Reading this book would be a good start for any research.

92 This work appeared in an article by Robert N. Miller, PhD. The Energies of Spiritual Healing, Science of Mind, January 1976, 22-27. See https://scienceofmindarchives. com/product/science-of-mind-magazine-01-january-1976/, sourced on 24 October 2021. These experiments are also mentioned in Chapter 33 (pages 431-444) of White, Future Science.

93 An interesting story depicting the more human interests surrounding this scientific concept can be found at http://tofspot.blogspot.com/2013/08/the-great-ptolemaic-smackdown.html, accessed on 12 June 2021.

94 Besant and Leadbeater, Occult Chemistry, 14.

95 Besant and Leadbeater, Occult Chemistry, 15.

96 Brooks, Quest.

97 Astral vision. See Leadbeater, The Astral Plane, 22-23.

98 Keen, Montague, The Scole Report.

99 Coleman, Tim. The Afterlife Investigations.

100 3D wave simulator—One that I found is at http://falstad. com/wavebox/ sourced on 17 March 2017. It allowed me to see radiation patterns from different antenna types and how changing the distance between dipoles and quadrupoles changed the shape of the radiation patterns. This tool also helped me realize the significance of the increasing distance between quadrupole elements in the higher electron shells.

101 This image was sourced from the article Stodolna, Hydrogen Atoms…, 213001-3.

102 Occult Chemistry corroborates that the sphere wall around the atoms is "at a great distance from the central group and is generally a sphere." It also says that there are a few exceptions, such as in nitrogen, where the enclosing spherical wall is an ovoid shape. If nitrogen atoms could be photographed, and they were shown to be an ovoid shape, would this be corroboration?

103 Besant and Leadbeater, Occult Chemistry, 2.

104 Besant and Leadbeater, Occult Chemistry, 15.

105 See Schlitz, Living Deeply. This is a wonderful book. It is the result of a ten-year study of transformations in human consciousness, which was conducted at the Institute for Noetic Sciences.

106 The icon titled Stand in the Mandorla is a real wooden icon that was created by artist Gaylene A. Barnes in New Zealand. It appears on the cover of my first three books. An interesting factoid, the photo of the icon is a composite of five separate shots that were stitched together, due to the difficulty of photographing the icon's shiny gold leaf surface.

107 May, The Courage to Create, 16–17.

108 May, The Courage to Create, 17.

109 Leadbeater, The Astral Plane, 10.

110 Leadbeater, The Astral Plane, 126.

111 Ziewe, Multidimensional Man, 69–70.

REFERENCE QR CODES

Use your mobile device camera to
convert each QR code into a clickable URL.

The American Center for
the Integration of Spiritually
Transformative Experiences

Bengston Research

Brian Faulkner
Content Editor

Earth's Hidden Reality

Earth's Hidden Reality
YouTube Channel

The Henry T. Laurency
Publishing Foundation

International Remote
Viewing Association

Lee Bladon's
Evolving Souls

Lynne McTaggart

MaryDes Designs
Cover Design

The Search for Higher
Dimensional Intelligence

SPARK Publications
Design, Publisher

Spiritual Science
Founders Association

The Theosophical
Society in America

Endnote #6

Endnote #8

Endnote #13

Endnote #14

Endnote #15

Endnote #16

Endnote #17

Endnote #18

Endnote #28

Endnote #29

Endnote #36

Endnote #37

Endnote #41

Endnote #44

Endnote #46

Endnote #48

Endnote #49
(iOS Phones only)

Endnote #54

Endnote #70

Endnote #86

Endnote #92

Endnote #93

Endnote #100

Visit **www.EarthsHiddenReality.com**

to connect with Mark Brooks and start a dialogue.